to Milo and Lia

LONDON MATHEMATICAL SOCIETY LECTURE NOTE SERIES

Managing Editor: Professor M. Reid, Mathematics Institute, University of Warwick, Coventry CV4 7AL, United Kingdom

The titles below are available from booksellers, or from Cambridge University Press at www.cambridge.org/mathematics

London Mathematical Society Lecture Note Series: 361

Words

Notes on Verbal Width in Groups

DAN SEGAL

All Souls College, Oxford

CAMBRIDGE
UNIVERSITY PRESS

CAMBRIDGE UNIVERSITY PRESS
Cambridge, New York, Melbourne, Madrid, Cape Town, Singapore, São Paulo, Delhi

Cambridge University Press
The Edinburgh Building, Cambridge CB2 8RU, UK

Published in the United States of America by Cambridge University Press, New York

www.cambridge.org
Information on this title: www.cambridge.org/9780521747660

© D. Segal 2009

First published 2009

Printed in the United Kingdom at the University Press, Cambridge

A catalogue record for this publication is available from the British Library

ISBN 978-0-521-74766-0 paperback

100600 1793

Contents

Preface

'Something old, something new, something borrowed ...'

One of Frobenius's many good ideas was to use character theory to study the solution of equations in groups. For example, he gave a formula for the number of solutions in a finite group G of equations like

$$[x, y] = g \tag{1}$$

(here $g \in G$ is a given element and the unknowns x, y range over G). After a long gap, these methods were taken up and applied with devastating effect to finite simple groups; recent highlights are Shalev's proof that for any non-trivial group word w, every element of every sufficiently large finite simple group is a product of three w-values, and the proof (by Liebeck, O'Brien, Shalev and Tiep) of 'Ore's Conjecture': Equation (1) is solvable for *every* element g in *every* finite simple group (of course 'simple' here means 'non-abelian simple').

This – very interesting – story needs to be told in another book, though I will survey some of the results in §4.6. My own interest in words began with Serre's well-known proof that in a finitely generated pro-p group, every subgroup of finite index is open. This rests on a (fairly elementary) result due to Peter Stroud: in a d-generator nilpotent group, every element of the derived group is a product of d commutators. A purely algebraic fact, this holds for groups of any size; but for Serre's theorem one only applies it to finite p-groups. Trying to generalize Serre's theorem to prosoluble groups, I had to establish a result like Stroud's that would hold in all finite soluble groups, and the only way I could find to do this was to use finiteness in a strong way (counting arguments). The restriction to finite groups became more obviously inevitable when Nikolay Nikolov and I finally extended Serre's theorem to all finitely generated profinite groups: the proof depends crucially on the classification of finite simple groups. It was also clear that on moving from soluble groups to finite groups in general we would have to consider group words other than commutators.

We were thus led to questions of the following type: given a (suitable) group word w, is there a function f on $\mathbb{N}$ such that the width of w in any finite group G is bounded by $f(\mathrm{d}(G))$? Here, $\mathrm{d}(G)$ denotes the minimal size of a generating set for G, and the *width* of w in G is the least integer n such that every product of w-values or their inverses in G is a product of n w-values or their inverses.

(Actually in the text I will use 'width' in a slightly wider sense: 'w has width n' will mean that *the* width of w is at most n.) If such a function f exists, I say that the word w is 'uniformly elliptic in finite groups'.

Various answers to this question are presented in Chapter 4, which also explores the connection with profinite groups. This is based on the observation (first articulated by Brian Hartley, as far as I know) that in a profinite group G, a word w has finite width if and only if the verbal subgroup $w(G)$ is *closed*; the bridge between algebra and topology exploited by Serre in the work referred to above.

Of course, these questions make perfectly good sense also for infinite groups, and in fact have a respectable history. In the 1960s Philip Hall studied various properties of word-values in groups, and introduced some characteristic terminology – a group G being called *w-elliptic* if w has finite width in G, and *verbally elliptic* if it is w-elliptic for every word w. One of the deepest results to emerge from Hall's school was the theorem that *every finitely generated abelian-by-nilpotent group is verbally elliptic*. This was obtained by Hall's student Peter Stroud, who was tragically killed in an accident shortly after completing his doctorate; as a result, the proof was never published, and remains quite difficult to get at (Stroud's thesis may be consulted in the Cambridge University Library, but photocopying is not allowed). Having tried and failed to reconstruct the proof myself, I thought it would be worthwhile to make the techniques more widely accessible. This idea was given a boost when Jim Roseblade kindly gave me a copy of another unpublished thesis, by Keith George: in this thesis, Stroud's results are generalized and his proof simplified. Combining the methods of Keith George and some ideas from a creative but rather condensed paper of V. A. Romankov, one is led to some quite general ellipticity results about infinite soluble groups. These are stated and proved in Chapter 2.

Chapter 1 and much of Chapter 2 are devoted to a systematic exposition of the relevant – mostly quite elementary – group-theoretic techniques.

The short Chapter 3 completes the picture by establishing that free groups are never w-elliptic (except in trivial cases), a result due to another of Hall's students, Akbar Rhemtulla.

The final Chapter 5 considers verbal ellipticity in algebraic groups, in certain profinite groups associated with these, and in p-adic analytic groups.

The heart of the work is in Chapter 4. The mathematics here is of necessity at a more sophisticated level than in the earlier parts, and not all proofs can be given in full. I have tried to give an accessible introduction to the recent work of Nikolov and myself concerning the width of words in finite and profinite groups, and an almost self-contained account of Andrei Jaikin's definitive results about verbal width in finite p-groups and pro-p groups. These two strands are then combined in §4.8, where I attempt to answer the question: *which words are uniformly elliptic in finite groups?* While the complete answer is still not known, the results obtained here are suggestive.

The book is both an exposition of a body of techniques in group theory and a report on work still in progress. I have tried to include everything that is known on the (rather specialized) topic[1], and to motivate a small number of key open problems (which are collected together in the appendix). There is a sprinkling of exercises, all supposed to be easy if one reads the *hints*.

Some of the results have been published (theorems of Rhemtulla, Merzljakov, Romankov, Nikolov-Segal and Jaikin). Some are not new but are previously unpublished (theorems of Stroud and George). A fair number are new; these include the verbal ellipticity of virtually soluble minimax groups in §2.6, that of certain adelic groups (§5.2), and the characterizations of uniformly elliptic words in §4.8. The material should therefore be of some interest to experts, while the expository style will, I hope, make it readily accessible to beginning researchers in the area.

Acknowledgements. I have lectured on parts of this material in Oxford and in Padova, and am grateful to the attentive audiences in both places. I am indebted to Gustavo Fernandez-Alcober and Luis Ribes for their careful reading of the text and a number of corrections. The elegant commutator identities in Exercises 1.2.2 and 1.2.3 are due to Gustavo Fernandez-Alcober.

There would be less of a story to tell but for the brilliant work of my two younger colleagues Nikolay Nikolov and Andrei Jaikin. Particular thanks also to Andrei for reading a draft of these notes and making several astute suggestions.

Copy-editing and proof-reading were meticulously carried out by Dave Edwards; remaining peculiarities are due to my own waywardness.

---

[1] This does not include the many interesting results on commutator width, see e.g. [DV], and 'stable commutator length', see [C].

Chapter 1

Generalities

1.1 Basic concepts

A *word* is an expression of the form

$$w(x_1, \ldots, x_k) = \prod_{j=1}^{s} x_{i_j}^{\varepsilon_j}$$

where $i_1, \ldots, i_s \in [1, k] := \{1, \ldots, k\}$ and each ε_j is ± 1. The *length* of w is s (this can be 0, when w is the 'empty word'). I will keep the symbol k for the number of variables, but this is not supposed to imply that every variable actually occurs in a given word (in practice, we may as well assume that k is a fixed, finite, but very large number). For any group G we have the *verbal mapping*

$$w : G^{(k)} \to G$$

$$\mathbf{g} \mapsto w(\mathbf{g}) = \prod_{i=1}^{s} g_{i_j}^{\varepsilon_j};$$

here and throughout I write

$$\mathbf{g} = (g_1, \ldots, g_k)$$

(and analogously with other letters of course); and $G^{(k)} = G \times \cdots \times G$ with k factors (to avoid confusion with the *subgroup* G^k generated by kth powers in G).

It is sometimes convenient to identify a word with an element of the free group F_k on $\{x_1, \ldots, x_k\}$. Different words may represent the same element of F_k, but of course they all induce the same verbal mapping. Indeed,

$$w(g_1, \ldots, g_k) = w\pi_{\mathbf{g}}$$

1

where $\pi_{\mathbf{g}}$ is the unique homomorphism $F_k \to G$ sending x_i to g_i for each i. This is discussed in a slightly more general setting in §1.3.

Let w be a word and G a group. We write

$$G_w = \left\{ w(\mathbf{g})^{\pm 1} \mid \mathbf{g} \in G^{(k)} \right\};$$

this is the (symmetrized) set of *w-values* in G.

For any subset S of G and $m \in \mathbb{N}$ we write

$$S^{*m} = \{ s_1 s_2 \ldots s_m \mid s_1, s_2, \ldots, s_m \in S \},$$

and denote by $\langle S \rangle$ the subgroup of G generated by S. The *verbal subgroup* corresponding to w is

$$w(G) = \langle G_w \rangle.$$

I will say that w has *width m* in G if

$$w(G) = G_w^{*m}.$$

Note: the word 'width' is here used in the wide sense, so 'width m' implies 'width n' for every $n \geq m$; we may define *the* width of w to be the least such m. One says that w has *infinite width* in G if it does not have finite width.

The group G is said to be *verbally elliptic* if every word has finite width in G.

Derived words

For $\mathbf{a}, \mathbf{g} \in G^{(k)}$ we set

$$w_{\mathbf{g}}'(\mathbf{a}) = w(\mathbf{a}.\mathbf{g}) w(\mathbf{g})^{-1}$$

where $\mathbf{a}.\mathbf{g} = (a_1 g_1, \ldots, a_k g_k)$. For $H \leq G$ and $Y \subseteq G$ write

$$w_Y'(H) = \left\langle w_{\mathbf{y}}'(\mathbf{a}) \mid \mathbf{a} \in H^{(k)}, \ \mathbf{y} \in Y^{(k)} \right\rangle.$$

Lemma 1.1.1 *If $H \triangleleft G = HY$ for some subset Y of G then*

$$w_Y'(H) = w_G'(H) \triangleleft G.$$

Proof. $w_G'(H)$ is normal in G because $w_{\mathbf{g}}'(\mathbf{a})^x = w_{\mathbf{g}^x}'(\mathbf{a}^x)$. Suppose $\mathbf{g} = \mathbf{b}.\mathbf{y}$ with $\mathbf{b} \in H^{(k)}$ and $\mathbf{y} \in Y^{(k)}$. Then for $\mathbf{a} \in H^{(k)}$ we have

$$
\begin{aligned}
w_{\mathbf{g}}'(\mathbf{a}) &= w(\mathbf{a}.\mathbf{b}.\mathbf{y}) w(\mathbf{b}.\mathbf{y})^{-1} \\
&= w_{\mathbf{y}}'(\mathbf{a}.\mathbf{b}) w(\mathbf{y}).w(\mathbf{y})^{-1} w_{\mathbf{y}}'(\mathbf{b})^{-1} \\
&= w_{\mathbf{y}}'(\mathbf{a}.\mathbf{b}) w_{\mathbf{y}}'(\mathbf{b})^{-1} \in w_Y'(H).
\end{aligned}
$$

∎

The marginal subgroup

This is

$$w^*(G) = \left\{ a \in G \mid w(\underline{a}^{(i)}.\mathbf{g}) = w(\mathbf{g}) \; \forall \mathbf{g} \in G^{(k)}, \; i = 1, \ldots, k \right\}$$

where

$$\underline{a}^{(i)} = (1, \ldots, a, \ldots, 1)$$

with a in the ith place. It is easily seen that $w^*(G)$ is a characteristic subgroup of G. A subgroup H of G is *marginal* for w if $H \le w^*(G)$.

It is clear that if $\mathbf{a} \in w^*(G)^{(k)}$ then $w(\mathbf{a}.\mathbf{g}) = w(\mathbf{g})$ for every $\mathbf{g}$. If $a \in w^*(G)$ then also $b_i = [a, g_i^{-1}] \in w^*(G)$ for $i = 1, \ldots, k$, and we have

$$w(\mathbf{g})^a = w(g_1^a, \ldots, g_k^a) = w(\mathbf{b}.\mathbf{g}) = w(\mathbf{g}),$$

whence the important observation:

$$[w(G), w^*(G)] = 1. \tag{1.1}$$

(Actually the argument shows that $[w(G), M] = 1$ where $M/w^*(G) = Z(G/w^*(G))$. A similar argument shows that the left-right asymmetry in the definition of $w^*(G)$ is only apparent.)

The following basic lemma implies that every word has finite width in a finite group:

Lemma 1.1.2 *If $G = \langle S \rangle$, $1 \in S$ and $|G| = n > 1$ then $G = S^{*(n-1)}$.*

Proof. Since G is finite, every element of G can be written in the form $g = s_1 \ldots s_m$ with each $s_j \in S$. If $m \ge n$ then two of the 'initial segments'

$$1, \; s_1, \; s_1 s_2, \; \ldots, \; s_1 \ldots s_m$$

must be equal; replacing the longer by the shorter we get a shorter expression for g. ∎

Proposition 1.1.3 *Suppose that K is a finite normal subgroup of G. If w has finite width in G/K then w has finite width in G.*

Proof. Put $W = w(G)$. Then

$$WK/K = w(G/K) = (G/K)_w^{*m} = G_w^{*m} K/K$$

where w has width m in G/K, whence

$$W = G_w^{*m}(K \cap W).$$

As $K \cap W$ is finite, there exists $r < \infty$ such that each element of $K \cap W$ is a product of at most r elements of G_w. Then $W = G_w^{*(m+r)}$. ∎

Exercise: Show that $K \cap W$ has a generating set contained in $G_w^{*(2m+1)}$, and deduce that w has width
$$(2m+1)\,|K|$$
in G.

Applying the above proposition to a quotient we get

Corollary 1.1.4 *Let $T < K$ be normal subgroups of G with K/T finite and $T \subseteq G_w^{*n}$ for some n. If w has finite width in G/K then w has finite width in G.*

Concatenation

Let $w_1, \ldots, w_t$ be words in k variables, or functions $G^{(k)} \to G$. Then
$$w_1 \divideontimes \cdots \divideontimes w_t$$
is the word, or function, v in tk variables given by
$$v(x_1, \ldots, x_{tk}) = \prod_{i=0}^{t-1} w_{i+1}(x_{ik+1}, \ldots, x_{ik+k}).$$
Note that then $v(G) = \langle w_1(G), \ldots, w_t(G) \rangle$ and that
$$G_v \subseteq G_{w_1} \cdot G_{w_2} \cdot \ldots \cdot G_{w_t} \cup G_{w_t} \cdot G_{w_{t-1}} \cdot \ldots \cdot G_{w_1} \subseteq G_v^{*t}.$$

(When the w_i are words, the set $G_{w_1} \cdot G_{w_2} \cdot \ldots \cdot G_{w_t}$ is invariant under permutations of the factors, since each G_{w_i} is invariant under conjugation; but this is no longer true for the 'generalized word functions' we will meet in §1.3 below.)

Some notation

$$H \le_f G, \ H \triangleleft_f G$$
means 'H (is) a subgroup (respectively normal subgroup) of finite index in G'.

$$F_k = F(x_1, \ldots, x_k), \ F_\infty = F(x_1, x_2, \ldots)$$
denote the free group on k, respectively countably infinitely many, free generators $x_1, x_2, \ldots$

Certain important words have a standard notation: the *commutator* of x and y is
$$\gamma_2(x, y) = [x, y] = x^{-1}y^{-1}xy.$$
For $n > 2$, the *left-normed repeated commutator* in n variables is
$$\gamma_n(x_1, \ldots, x_n) = [x_1, \ldots, x_n] = [\gamma_{n-1}(x_1, \ldots, x_{n-1}), x_n].$$

For subgroups $A_1, A_2, \ldots, A, B$ of a group G,

$$[A, B] = [A,_1 B] = \langle [a, b] \mid a \in A, \ b \in B \rangle,$$
$$[A_1, A_2, \ldots, A_n] = [[A_1, A_2, \ldots, A_{n-1}], A_n],$$
$$[A,_n B] = [[A,_{n-1} B], B].$$

But for $g \in G$ and $S \subseteq G$ we write

$$[S, g] = \{[s, g] \mid s \in S\}$$
$$[g, S] = \{[g, s] \mid s \in S\}$$

(each is a *set* of values, not the subgroup they generate).

The verbal subgroups corresponding to the commutator words are

$$G' := \gamma_2(G) = [G, G],$$

the derived group, and for $n > 2$

$$\gamma_n(G) = \left\langle [g_1, \ldots, g_n] \mid \mathbf{g} \in G^{(n)} \right\rangle$$

(*note*: this is not the usual definition! But don't worry, see Exercise 1.2.1 below).

The term **rank** is always used in the sense of *Prüfer rank*; that is

$$\mathrm{rk}(G) = \sup\{\mathrm{d}(H) \mid H \text{ is a finitely generated subgroup of } G\}$$

and $\mathrm{d}(H) \in \mathbb{N} \cup \{\infty\}$ denotes the minimal size of a generating set for H.

1.2 Commutators

The commutator

$$\gamma_2(x, y) = [x, y] = x^{-1}y^{-1}xy$$

is the simplest really interesting word. A good understanding of how it behaves is both the first step to understanding more complicated words, and an essential tool in many arguments in group theory. The key feature of γ_2 as a word mapping is that it is 'bilinear to a first approximation'; this simple observation has a surprising amount of mileage in it (it may be seen as the foundation of Lie theory, for example). Here we record some of its more or less direct consequences. Most of the basic results will be taken for granted in later sections.

Let's start with

Exercise 1.2.1. Prove that if $A_1, A_2, \ldots, A_n$ are normal subgroups of G then

$$[A_1, A_2, \ldots, A_n] = \langle [a_1, \ldots, a_n] \mid a_j \in A_j \ (j = 1, \ldots, n) \rangle.$$

Deduce that the verbal subgroup for the word γ_n is indeed the nth term of the lower central series of G, namely

$$\gamma_n(G) = [G, _{n-1} G]$$

(so our verbal subgroup notation is consistent with the usual one for the lower central series).

This exercise is a typical application of the following basic identities. These express the bilinearity of γ_2 'modulo higher commutators'; they will frequently be used without special mention.

$$[x,y]^{-1} = [y,x] = [x^y, y^{-1}] = [x, y^{-1}][x, y^{-1}, y]$$
$$[x^{-1}, y] = [x, y^{x^{-1}}]^{-1} = [x, y, x^{-1}]^{-1}[x, y]^{-1}$$
$$[xy, z] = [x, z]^y[y, z] = [x^y, z^y][y, z] = [x, z][x, z, y][y, z]$$
$$[x, yz] = [x, z][x, y]^z = [x, z][x^z, y^z] = [x, z][x, y][x, y, z].$$

Another useful identity is the *Hall-Witt identity*, the group-theoretic analogue of the Jacobi identity:

$$[x, y^{-1}, z]^y[y, z^{-1}, x]^z[z, x^{-1}, y]^x = 1.$$

From this one deduces the **Three-subgroup Lemma**: if X, Y and Z are normal subgroups of G then

$$[X, Y, Z] \leq [Y, Z, X][Z, X, Y];$$

more generally, if X, Y and Z are any subgroups of G then $[X, Y, Z]$ is contained in the normal closure of $[Y, Z, X][Z, X, Y]$.

Arguing by induction on n, it is easy to deduce a fundamental property of the lower central series:

$$[\gamma_m(G), \gamma_n(G)] \leq \gamma_{m+n}(G). \tag{1.2}$$

The multilinear nature of the commutator mapping is expressed in

Proposition 1.2.1 *Let $n \geq 2$ and let $1 \leq i \leq n$. Fix $x_1, \ldots, x_n \in G$ and define $f_i = f_i^n : G \to \gamma_n(G)$ by*

$$f_i(g) = [x_1, \ldots, x_{i-1}, g, x_{i+1}, \ldots, x_n].$$

If $x \in G$ and $y \in \gamma_m(G)$ then

$$f_i(xy) \equiv f_i(x)f_i(y) \mod \gamma_{n+m}(G) \tag{1.3}$$
$$\equiv f_i(x) \mod \gamma_{n+m-1}(G). \tag{1.4}$$

Proof. Write $G_s = \gamma_s(G)$ for each s. Note that (1.4) follows from (1.3) and (1.2). We first prove (1.3) for $i = 1$, by induction on n. When $n = 2$ the claim is

$$[xy, z] \equiv [x, z][y, z] \mod G_{2+m};$$

this is clear from the basic identities. Suppose $n > 2$, and assume inductively that

$$f_1^{n-1}(xy) = f_1^{n-1}(x)f_1^{n-1}(y)z$$

with $z \in G_{n-1+m}$. Then

$$\begin{aligned}
f_1^n(xy) &= [f_1^{n-1}(xy), x_n] \\
&= [f_1^{n-1}(x)f_1^{n-1}(y)z, x_n] \\
&\equiv f_1^n(x)f_1^n(y) \mod G_{n+m},
\end{aligned}$$

again using some basic identities.

Now suppose that $i > 1$, and put $u = [x_1, \ldots, x_{i-1}]$. Then

$$\begin{aligned}
f_i^n(xy) &= [u, xy, x_{i+1}, \ldots, x_n] \\
&= [[u, x][u, y]z, x_{i+1}, \ldots, x_n] \\
&= \widetilde{f}_1^{n-i+1}([u, x][u, y]z)
\end{aligned}$$

where $z = [[u, y], [u, x]][u, x, y] \in G_{i+m}$, and $\widetilde{f}_1^{n-i+1}$ is obtained from f_1^{n-i+1} on replacing each x_j by x_{i+j-1}. Since $[u, y]z \in G_{i-1+m}$, two applications of the first part with $\widetilde{f}_1$ in place of f_1 show that

$$f_i^n(xy) \equiv \widetilde{f}_1^{n-i+1}([u, x])\widetilde{f}_1^{n-i+1}([u, y])\widetilde{f}_1^{n-i+1}(z) \mod G_{n+m}.$$

The result follows since the first two factors are equal, respectively, to $f_i^n(x)$ and $f_i^n(y)$, and the third factor lies in G_{n+m}. (For a slicker and more general proof see Exercise 1.2.2 below.) ∎

Corollary 1.2.2 *Let G be a nilpotent group of class $c \geq n$. Then $\gamma_{c-n+2}(G)$ is marginal for γ_n in G.*

Corollary 1.2.3 *Let G be a nilpotent group of class $c > 1$. Then γ_c induces a multilinear mapping*

$$(G/G')^{(c)} \to \gamma_c(G).$$

Corollary 1.2.4 *Let $w \in \gamma_n(\eta(F_k))$ where η is any word. Let H be any group. Then $\gamma_m(\eta(H))$ is marginal for w in H modulo $\gamma_{n+m-1}(\eta(H))$.*

Proof. Write $F = F_k$. We have

$$w = \prod_j \gamma_n(u_{j1}, \ldots, u_{jn})^{\varepsilon_j}$$

where each $u_{jl} \in \eta(F)$ is a word in $x_1, \ldots, x_k$ and $\varepsilon_j = \pm 1$. Fix $i \in \{1, \ldots, n\}$, let $y \in \gamma_m(\eta(F))$ and put

$$v_{jl} = u_{jl}(x_1, \ldots, x_{i-1}, x_i y, x_{i+1}, \ldots, x_n).$$

Then
$$v_{jl} \equiv u_{ji} \mod \gamma_m(\eta(F))$$
for each j and l. Hence by Proposition 1.2.1 (applied with $G = \eta(F)$) we have
$$\gamma_n(v_{j1}, \ldots, v_{jn}) \equiv \gamma_n(u_{j1}, \ldots, u_{jn}) \mod \gamma_{n+m-1}(\eta(F)).$$
Therefore
$$w(x_1, \ldots, x_{i-1}, x_i y, x_{i+1}, \ldots, x_n) \equiv w(x_1, \ldots, x_{i-1}, x_i, x_{i+1}, \ldots, x_n) = w$$
modulo $\gamma_{n+m-1}(\eta(F))$, and the result follows on evaluating this in H. ∎

A different kind of application is:

Proposition 1.2.5 *Let G be a group, H a normal subgroup and suppose that $G = G' \langle x_1, \ldots, x_m \rangle$. Then*
$$[H, G] = [H, x_1] \ldots [H, x_m][H,_n G]$$
for every $n \geq 1$.

Proof. Suppose this holds for a certain value of $n \geq 1$. To deduce that it holds with $n + 1$ in place of n we may as well assume that $[H,_{n+1} G] = 1$. Now put $A = [H,_{n-1} G]$. Then $(a, g) \mapsto [a, g]$ induces a bilinear mapping
$$A \times G/G' \to [H,_n G] = B \leq \mathrm{Z}(G).$$
It follows that
$$B = [A, x_1] \ldots [A, x_m].$$
With the inductive hypothesis this implies that
$$\begin{aligned}
[H, G] &= [H, x_1] \ldots [H, x_m] B \\
&= [H, x_1] \ldots [H, x_m][A, x_1] \ldots [A, x_m] \\
&= \prod_{i=1}^{m} [H, x_i][A, x_i] \\
&= \prod_{i=1}^{m} [H, x_i];
\end{aligned}$$
the third equality holds because each $[A, x_i] \subseteq \mathrm{Z}(G)$, and the final equality holds because
$$[h, x][a, x] \equiv [ha, x] \mod[H, G, A] = 1$$
for $h \in H$, $a \in A$ and $x \in G$. ∎

If G is nilpotent of class c, we can take $n \geq c$ and $H = G$ to infer

Corollary 1.2.6 *Suppose that $G = G' \langle x_1, \ldots, x_m \rangle$ is nilpotent. Then*
$$\gamma_2(G) = [G, x_1] \ldots [G, x_m].$$

It follows that *the word γ_2 has width m in every m-generator nilpotent group.*
A similar result holds when G is a *finite* soluble group (Exercise 4.7.4), but the
following is open:

Problem 1.2.1 Does γ_2 have finite width in every finitely generated soluble
group?

In [S11] Stroud speculates that there may be a counterexample G satisfying
$[G''', G] = 1$.

More generally, we have

Proposition 1.2.7 *If $G = G' \langle x_1, \ldots, x_m \rangle$ is nilpotent and $t, r \in \mathbb{N}$ then*

$$\gamma_{t+r}(G) = \prod [\gamma_r(G), x_{i_1}, \ldots, x_{i_t}],$$

the product ranging over $\mathbf{i} = (i_1, \ldots, i_t) \in [1, m]^{(t)}$, *in any chosen order.*

Proof. Write $G_n = \gamma_n(G)$ for each n. First we prove by induction on t that for
$s \in \mathbb{N}$,

$$G_{t+s} = \prod_{\mathbf{i}} [G_s, x_{i_1}, \ldots, x_{i_t}] \cdot G_{t+s+1}. \tag{1.5}$$

If $t = 1$ this follows from Proposition 1.2.5 with $H = G_s$. Now let $t > 1$ and
take $H = G_{t-1+s}$. Then

$$\begin{aligned}
[H, G] &= \prod_{j=1}^{m} [H, x_j] \cdot G_{t+s+1} \\
&= \prod_{j=1}^{m} \left[\prod_{\mathbf{i}} [G_s, x_{i_1}, \ldots, x_{i_{t-1}}] \cdot G_{t+s}, x_j \right] \cdot G_{t+s+1} \\
&= \prod_{\mathbf{j}} \left[[G_s, x_{i_1}, \ldots, x_{i_{t-1}}], x_j \right] \cdot G_{t+s+1},
\end{aligned}$$

using the inductive hypothesis at the second step and Proposition 1.2.1 at the
third step. This establishes (1.5).

Now let $n \geq 1$ and suppose that

$$G_{t+s} = \prod_{\mathbf{i}} [G_s, x_{i_1}, \ldots, x_{i_t}] \cdot G_{t+s+n} \tag{1.6}$$

for every $s \in \mathbb{N}$. Taking $s = r$ and $s = r + n$ we get

$$\begin{aligned}
G_{t+r} &= \prod_{\mathbf{i}} [G_r, x_{i_1}, \ldots, x_{i_t}] \cdot \prod_{\mathbf{i}} [G_{r+n}, x_{i_1}, \ldots, x_{i_t}] \cdot G_{t+r+n+1} \\
&= \prod_{\mathbf{i}} [G_r, x_{i_1}, \ldots, x_{i_t}] \cdot G_{t+r+n+1}
\end{aligned}$$

by Proposition 1.2.1. Thus by induction (1.6) holds for every n, and the result
follows since $G_{t+r+n} = 1$ for sufficiently large n. $\blacksquare$

Corollary 1.2.8 *If $G = G' \langle x_1, \ldots, x_m \rangle$ is nilpotent and $t \geq 1$ then*

$$\gamma_{t+1}(G) = \prod_{\mathbf{i} \in [1,m]^t} [G, x_{i_1}, \ldots, x_{i_t}].$$

The word γ_{t+1} has width m^t in G.

Remark. Recall that if G is a nilpotent group, we have $G = G' \langle x_1, \ldots, x_m \rangle$ if and only if $G = \langle x_1, \ldots, x_m \rangle$ – e.g. because the subgroup $\langle x_1, \ldots, x_m \rangle$ is in any case subnormal in G, so if it is proper it is contained in some proper normal subgroup N, and then $G > G'N \geq G' \langle x_1, \ldots, x_m \rangle$.

Exercise 1.2.2. (i) Let U, A and B be normal subgroups of a group G and for $n \geq 0$ set

$$Q_n = \prod_{r+s=n} [A,_r G, B,_s G], \quad P_n = \prod_{r+s=n} [U, A,_r G, B,_s G]$$

(reading $[X,_0 Y] = X$). Prove that if $u \in U$, $a \in A$, $b \in B$ and $y_1, \ldots, y_n \in G$ then

$$[ab, y_1, \ldots, y_n] \equiv [a, y_1, \ldots, y_n][b, y_1, \ldots, y_n] \mod Q_n$$

and

$$[u, ab, y_1, \ldots, y_n] \equiv [u, a, y_1, \ldots, y_n][u, b, y_1, \ldots, y_n] \mod P_n.$$

[*Hint*: Both parts are proved the same way, using induction on n.]
(ii) Deduce Proposition 1.2.1.

Exercise 1.2.3. Let G be a group, H a normal subgroup and suppose that $G = G' \langle x_1, \ldots, x_m \rangle$. Prove that

$$[H,_t G] = \prod [H, x_{i_1}, \ldots, x_{i_t}] \cdot [H,_l G]$$

for every $l \geq t$, the product ranging over $\mathbf{i} = (i_1, \ldots, i_t) \in [1, m]^{(t)}$, in any chosen order. [*Hint*: assume without loss of generality that $[H,_l G] = 1$, and argue by induction on $l - t$, using the preceding exercise.]

1.3   Generalized words

It is usual to consider words as elements of a free group. We shall also need to deal with 'generalized words', that is, words twisted by group automorphisms. These are conveniently identified with elements of a free 'group with operators'. Let $X = \{x_1, \ldots, x_k\}$ and $\Phi = \{\phi_1, \ldots, \phi_s\}$ be disjoint finite alphabets, and set

$$F_\Phi(X) = \langle x^\phi \mid x \in X, \ \phi \in \Phi \rangle < F(X \cup \Phi)$$

where $F(X \cup \Phi)$ is the free group on $X \cup \Phi$. It is easy to see that $F_\Phi(X)$ is actually free on the exhibited generating set (*Exercise*: consider its image under the natural homomorphism from $F(X \cup \Phi)$ onto the wreath product

$F(X) \wr F(\Phi)$). An element of $F_\Phi(X)$ will be called an *abstract generalized word* in k variables.

Let us change perspective. A *generalized word function* (of length s, in k variables) on a group G is a mapping $\xi : G^{(k)} \to G$ of the form

$$\mathbf{g} \longmapsto \xi(\mathbf{g}) = \prod_{j=1}^{s} g_{i_j}^{\varepsilon_j \alpha_j}$$

where $i_1, \ldots, i_s \in \{1, \ldots, k\}$, $\varepsilon_j = \pm 1$ and $\alpha_j \in \mathrm{Aut}(G)$ for each j. Now for each $\mathbf{g} \in G^{(k)}$ there is a unique homomorphism $\pi_{\mathbf{g}} : F_\Phi(X) \to G$ such that

$$x_i^{\phi_j} \pi_{\mathbf{g}} = g_i^{\alpha_j}$$

for every i and j. Taking

$$v = \prod_{j=1}^{s} (x_{i_j}^{\phi_j})^{\varepsilon_j} \in F_\Phi(X) \tag{1.7}$$

we have

$$\xi(\mathbf{g}) = v \pi_{\mathbf{g}}.$$

In this sense, each generalized word function on G is obtained from an abstract generalized word by specifying some automorphisms of G.

As for ordinary words, we set $G_\xi = \{\xi(\mathbf{g})^{\pm 1} \mid \mathbf{g} \in G^{(k)}\}$, $\xi(G) = \langle G_\xi \rangle$, and say that ξ has width m in G if $\xi(G) = G_\xi^{*m}$.

An ordinary group word w gives rise to a family of generalized words in the following way. Suppose $G \lhd E$ are groups. Then for each fixed $\mathbf{h} \in E^{(k)}$ we get a generalized word function ξ on G by setting

$$\xi(\mathbf{g}) = w'_{\mathbf{h}}(\mathbf{g}).$$

Indeed, if $w = \prod_{j=1}^{s} x_{i_j}^{\varepsilon_j}$ then

$$w'_{\mathbf{h}}(\mathbf{g}) = \prod_{j=1}^{s} (g_{i_j} h_{i_j})^{\varepsilon_j} \left(\prod_{j=1}^{s} h_{i_j}^{\varepsilon_j} \right)^{-1} = \prod_{j=1}^{s} g_{i_j}^{\varepsilon_j \alpha_j}$$

where each α_j is the restriction to G of a suitable inner automorphism of E, depending only on w and $\mathbf{h}$. In particular, ξ has the same length and involves the same variables as w. The evident relation

$$G_\xi \subseteq E_w^{*2}$$

will be used frequently.

Put $F = \langle X \rangle < F(X \cup \Phi)$. For $\mathbf{u} = (u_1, \ldots, u_k) \in F^{(k)}$ let $\theta_{\mathbf{u}}$ be the endomorphism of $F(X \cup \Phi)$ defined by

$$x_i \longmapsto u_i \ (i = 1, \ldots, k)$$
$$\phi \longmapsto \phi \ (\phi \in \Phi).$$

Then $\theta_{\mathbf{u}}$ maps $F_\Phi(X)$ into itself, and we can 'evaluate' an abstract generalized word $v \in F_\Phi(X)$ at $\mathbf{u}$ by writing $v(\mathbf{u}) = v\theta_{\mathbf{u}}$. (This comes to the same as considering v as a generalized word function on the normal closure of F in $F(X \cup \Phi)$, the elements of Φ acting by conjugation.) Write $F_v = \left\{ v(\mathbf{u})^{\pm 1} \mid \mathbf{u} \in F^{(k)} \right\}$, $v(F) = \langle F_v \rangle$.

If $\mathbf{w} \in F^{(k)}$ then $x_i \theta_{\mathbf{u}} \theta_{\mathbf{w}} = u_i \theta_{\mathbf{w}} = u_i(\mathbf{w})$ for each i; as $\theta_{\mathbf{u}}$ and $\theta_{\mathbf{w}}$ fix each $\phi \in \Phi$ it follows that

$$v(\mathbf{u})\theta_{\mathbf{w}} = v(u_1(\mathbf{w}), \ldots, u_k(\mathbf{w}))$$

– which is simply saying that substitution is an associative operation. Thus

$$F_v \theta_{\mathbf{w}} \subseteq \langle w_1, \ldots, w_k \rangle_v \subseteq F_v. \tag{1.8}$$

If v corresponds to a generalized word function ξ on a group G, as above, we have the following more or less obvious relation:

$$v(\mathbf{u})\pi_{\mathbf{g}} = v\theta_{\mathbf{u}}\pi_{\mathbf{g}} = \xi(u_1(\mathbf{g}), \ldots, u_k(\mathbf{g})) \tag{1.9}$$

for $\mathbf{u} = (u_1, \ldots, u_k) \in F^{(k)}$ and $\mathbf{g} \in G^{(k)}$.

Proposition 1.3.1 *Let $v \in F_\Phi(X)$ and $c \in \mathbb{N}$. Then there exists $f \in \mathbb{N}$ such that*

$$v(F) \subseteq F_v^{*f} \cdot \gamma_{c+1}(F_\Phi(X)).$$

Before proving this we deduce

Theorem 1.3.2 *Let G be a nilpotent group of class c and finite rank d, and let ξ be a generalized word function on G, of length s. Then ξ has finite width in G, bounded in terms of c, d and s.*

Proof. Certainly ξ involves at most s variables, and we set $k = \max\{d, s\}$. Let $g \in \xi(G)$; then $g \in \langle \xi(\mathbf{h}) \mid \mathbf{h} \in H^{(k)} \rangle$ for some finitely generated subgroup H of G. As G has rank $d \leq k$ we have $H = \langle b_1, \ldots, b_k \rangle$ for some $b_1, \ldots, b_k \in G$, and then

$$H^{(k)} = \left\{ (u_1(\mathbf{b}), \ldots, u_k(\mathbf{b})) \mid \mathbf{u} \in F^{(k)} \right\}.$$

Let $v \in F_\Phi(X)$ correspond to ξ as above. Then using (1.9) we have

$$\left\{ \xi(\mathbf{h})^{\pm 1} \mid \mathbf{h} \in H^{(k)} \right\} = \left\{ (v\theta_{\mathbf{u}}\pi_{\mathbf{b}})^{\pm 1} \mid \mathbf{u} \in F^{(k)} \right\}$$
$$= F_v \pi_{\mathbf{b}}.$$

Thus

$$g \in \langle F_v \pi_{\mathbf{b}} \rangle$$
$$= v(F)\pi_{\mathbf{b}} \subseteq (F_v \pi_{\mathbf{b}})^{*f} \cdot \gamma_{c+1}(F_\Phi(X)\pi_{\mathbf{b}})$$
$$\subseteq G_\xi^{*f}$$

by Proposition 1.3.1. Here f depends on v, but there are only finitely many possibilities for v (in fact $(2k)^s$, as is clear from (1.7)). Thus f is bounded above in terms of the given parameters. ∎

We now proceed to the proof of Proposition 1.3.1. Let $H_i = \gamma_i(F_\Phi(X))$ for each $i \geq 1$.

For $n \in \mathbb{N}$ write

$$\theta_n = \theta_{(x_1^n, \ldots, x_k^n)}.$$

Thus θ_n maps $F_\Phi(X)$ into itself, raising each generator to the nth power.

Lemma 1.3.3 *If* $h \in H_i$ *then*

$$h\theta_n \equiv h^{n^i} \mod H_{i+1}.$$

Proof. This holds because H_i is generated modulo H_{i+1} by repeated commutators $\gamma_i(\mathbf{y}) = [y_1, \ldots, y_i]$ with each $y_j \in \{x^\phi \mid x \in X, \ \phi \in \Phi\}$, and the mapping induced by γ_i from $(H_1/H_2)^{(i)}$ into H_i/H_{i+1} is multilinear. ∎

We also need a result from number theory:

Theorem 1.3.4 ([HW], Theorem 401) *For each* $c \in \mathbb{N}$ *there exists* $\beta = \beta(c) \in \mathbb{N}$ *such that every integer is equal to a sum of* β *numbers each of which is* ± 1 *times the cth power of an integer.*

Proof. The $c + 1$ polynomials $f_j(X) = (X + j)^c$ $(j = 0, \ldots, c)$ are linearly independent over $\mathbb{Q}$. (*Exercise:* evaluate the determinant formed by the matrix of coefficients!) They therefore span the space of all polynomials of degree at most c, so in particular there exist rational numbers $q_0, \ldots, q_c$ such that

$$X = \sum_{j=0}^{c} q_j f_j(X).$$

Say $q_j = s_j/m$ with $m \in \mathbb{N}$ and $s_j \in \mathbb{Z}$. Then for $x \in \mathbb{Z}$ and $0 \leq r \leq m/2$ we have

$$mx \pm r = \sum_{j=0}^{c} s_j(x + j)^c \pm \sum_{j=1}^{r} 1^c.$$

So we can take $\beta(c) = \sum_0^c |s_j| + [m/2]$. ∎

(Note that this is quite elementary, unlike the celebrated 'Waring's Theorem' of Hilbert which says that every *positive* integer is the sum of boundedly many cth powers of *positive* integers.)

To prove Proposition 1.3.1 we argue by induction on c. If $c = 1$ we can take $f = 1$ since $\mathbf{u} \longmapsto v(\mathbf{u})H_2$ is a homomorphism $F^{(k)} \to H_1/H_2$. Now let $c \geq 1$ and suppose inductively that $v(F) \subseteq F_v^{*f_1} H_c$. As H_c/H_{c+1} is a finitely generated abelian group, there exist $h_1, \ldots, h_t \in v(F) \cap H_c$ such that

$$v(F) \cap H_c = \langle h_1 \rangle \ldots \langle h_t \rangle H_{c+1}.$$

There exists $e \in \mathbb{N}$ such that $h_j \in F_v^{*e}$ for each $j \in \{1, \ldots, t\}$. Fix j and put $h = h_j$. Given $n \in \mathbb{Z}$ there exist $n_1, \ldots, n_\beta \in \mathbb{Z}$ such that $n = \varepsilon_1 n_1^c + \cdots + \varepsilon_\beta n_\beta^c$ where $\beta = \beta(c)$ is given by Theorem 1.3.4 and each ε_j is ± 1. Then

$$h^n = \prod_{l=1}^{\beta} h^{\varepsilon_l n_l^c}$$

$$\equiv \prod_{l=1}^{\beta} (h\theta_{n_l})^{\varepsilon_l} \mod H_{c+1}.$$

As $h \in F_v^{*e}$, we have $h\theta_{n_l} \in F_v^{*e}$ for each l, by (1.8); it follows that $\langle h \rangle \subseteq F_v^{*e\beta} H_{c+1}$. Thus

$$v(F) \cap H_c \subseteq F_v^{*e\beta t} H_{c+1},$$

whence

$$v(F) = F_v^{*f_1}(v(F) \cap H_c) \subseteq F_v^{*f} H_{c+1}$$

where $f = f_1 + e\beta t$. This completes the proof.

The next result will be needed later, in §2.6.

Lemma 1.3.5 *Let G be a nilpotent group of class c, and let ξ be a generalized word function on G. Let $m \in \mathbb{N}$. Then*

$$\xi(\mathbf{g})^{p(m,c)} \in \xi(G^m)$$

for every $\mathbf{g} \in G^{(k)}$ where $p(m,c) = m^{c(c+1)/2}$.

Proof. Say ξ corresponds to the abstract generalized word v as above. I claim that

$$v^{p(m,c)} \in v(F)\theta_m \cdot H_{c+1}. \tag{1.10}$$

If this is true, we may deduce that

$$\xi(\mathbf{g})^{p(m,c)} = v^{p(m,c)}\pi_{\mathbf{g}} \in v(F)\theta_m \pi_{\mathbf{g}}$$
$$\leq v(\langle x_1^m, \ldots, x_k^m \rangle)\pi_{\mathbf{g}}$$

by (1.8), and it follows by (1.9) that

$$\xi(\mathbf{g})^{p(m,c)} \in \xi(\langle g_1^m, \dots, g_k^m \rangle) \le \xi(G^m).$$

To prove (1.10) we argue by induction on c. Suppose

$$v^{p(m,c-1)} = b\theta_m \cdot h$$

with $b \in v(F)$ and $h \in H_c$ (if $c = 1$ take $b = 1$ and $h = v$). Then using Lemma 1.3.3 we have

$$\begin{aligned}
v^{p(m,c)} &= (b\theta_m \cdot h)^{m^c} \\
&\equiv (b\theta_m)^{m^c} h^{m^c} \quad \mod H_{c+1} \\
&\equiv (b\theta_m)^{m^c} h\theta_m \quad \mod H_{c+1} \\
&= (b^{m^c} h)\theta_m.
\end{aligned}$$

Since θ_m maps $v(F)$ into itself, $h \in v(F)$, so $b^{m^c} h \in v(F)$ and (1.10) follows. ∎

1.4 Conciseness

The word w is said to be *concise* in a class of groups $\mathcal{C}$ if

$$|G_w| < \infty \implies |w(G)| < \infty$$

for every $G \in \mathcal{C}$. Note that this a 'local property' in the following sense: if w is concise in $\mathcal{C}$ and if G is *locally-$\mathcal{C}$* (i.e. every finite subset of G is contained in a $\mathcal{C}$-subgroup of G) then w is concise in $\mathcal{C} \cup \{G\}$ (for if G_w is finite then we can pick $H \le G$ with $H \in \mathcal{C}$ such that $G_w = H_w$).

Philip Hall conjectured that every word is concise in the class of all groups, but this is not true; a counterexample was constructed by S. Ivanov [I]. However, Jeremy Wilson [W2] proved that every outer-commutator word has this property. Apparently it is still an open problem whether Hall's conjecture holds in residually finite groups; the next lemma may be relevant to this.

Lemma 1.4.1 *Suppose G is a residually finite group and w is any word. Then G_w is finite if and only if $|G : w^*(G)|$ is finite.*

Proof. If G is any group and w is a word in k variables then clearly

$$|G_w| \le |G : w^*(G)|^k,$$

so G_w is finite if $|G : w^*(G)|$ is. For the converse, suppose that G_w is finite. Then so is the set

$$S = \left\{ w'_{\mathbf{g}}(\underline{a}^{(i)}) \mid \mathbf{g} \in G^{(k)}, \ a \in G, \ i = 1, \dots, k \right\}.$$

As G is residually finite, there is a normal subgroup N of finite index in G such that $N \cap S = \{1\}$. If $a \in N$ then for every $\mathbf{g} \in G^{(k)}$ we also have $w'_{\mathbf{g}}(\underline{a}^{(i)}) \in N$, whence $w'_{\mathbf{g}}(\underline{a}^{(i)}) = 1$. Thus $N \le w^*(G)$ and so $|G : w^*(G)| \le |G : N| < \infty$. ∎

Exercise. Show by example that residual finiteness is really needed here. [*Hint:* construct a class-2 nilpotent group with finite centre.]

Definition (i) A group G is in $(\mathrm{R}\mathfrak{F})^Q$ if every quotient of G is residually finite – that is, if every normal subgroup of G is closed in the profinite topology.

(ii) A group G is in $(\mathrm{R}\mathfrak{F})^*$ if for each $m \in \mathbb{N}$ there is a normal subgroup $K = K(m)$ of finite index in G such that K is residually (finite of order prime to m). Note that then the subgroups of finite m-prime index in K that are normal in G intersect in $\{1\}$; in particular this implies that G is residually finite.

The class $(\mathrm{R}\mathfrak{F})^Q$ contains all finitely generated virtually abelian-by-polycyclic groups (see [LR], §**7.2**). The class $(\mathrm{R}\mathfrak{F})^*$ contains all finitely generated linear groups in characteristic 0 ([W1], Theorem 4.7), all finitely generated groups that are virtually (torsion-free and abelian-by-polycyclic) [S2], and every virtually torsion-free virtually soluble minimax group ([LR], **5.3.9**).

Theorem 1.4.2 *Let $\mathcal{C}$ be the class of groups that are locally-$\left((\mathrm{R}\mathfrak{F})^Q \cup (\mathrm{R}\mathfrak{F})^*\right)$. Then every word is concise in $\mathcal{C}$.*

This combines results of Turner-Smith [TS] and Merzljakov [M3]. It depends on

Lemma 1.4.3 *Let $N \lhd G$ be finite groups and put $m = |G : N|$. Then $G = NH$ for some subgroup H with $|H|$ dividing some power of m.*

Proof. Suppose $G = NT$ for some proper subgroup T of G. Arguing by induction on $|G|$, we may suppose that $T = (N \cap T)H$ where $|H|$ divides some power of $|T : N \cap T| = m$. Then $G = NH$ and we are done.

Otherwise, N is contained in the Frattini subgroup of G, which is nilpotent. If $N = 1$ we take $H = G$. If $N > 1$, let M be a minimal normal subgroup of G contained in N; then M is an elementary abelian group of prime-power order p^e, say. Inductively, we may suppose that $G/M = (N/M)(L/M)$ where $|L/M| \mid m^{n_1}$ for some n_1; then $G = NL$ which now implies $L = G$. Thus $|G : M| \mid m^{n_1}$.

If $p \nmid m$ then G splits over M by the Schur-Zassenhaus theorem, which is impossible since M is contained in the Frattini subgroup of G. Therefore $p \mid m$. But then $|G| \mid p^e m^{n_1} \mid m^{n_1 + e}$ and we may take $H = G$. ∎

Proof of Theorem 1.4.2. It will suffice to show that if $G \in (\mathrm{R}\mathfrak{F})^Q \cup (\mathrm{R}\mathfrak{F})^*$ and w is a word such that G_w is finite then $w(G)$ is finite. Any such group is in particular residually finite, so we know from Lemma 1.4.1 that $w^*(G)$ has finite index m, say, in G.

(i) Suppose $G \in (\mathrm{R}\mathfrak{F})^Q$. Put $A = w(G) \cap w^*(G)$. This is an abelian group, in view of (1.1), and it is finitely generated because $w(G)$ is finitely generated and $|w(G) : A| \mid m < \infty$.

Choose a prime $p \nmid m$. Then G/A^p is residually finite, so G has a normal subgroup N of finite index with $N \cap A = A^p$. Applying Lemma 1.4.3 to $w^*(G)N/N \lhd G/N$, we find $H/N \leq G/N$ such that $G = w^*(G)H$ and $|H/N|$ divides some power of m. As AN/N is a p-group, we have $H \cap AN = N$. It follows that

$$A = w(G) \cap A = w(H) \cap A \leq H \cap A \leq N \cap A = A^p.$$

This is impossible if A is infinite (since in that case A has an infinite cyclic quotient). Therefore A is finite, and hence so is $w(G)$.

(ii) Suppose $G \in (\mathrm{R}\mathfrak{F})^*$, and let K be a normal subgroup of finite index that is residually (finite of order prime to m). Let $N \lhd G$ be a subgroup of finite m'-index in K. Applying Lemma 1.4.3 to $w^*(G)N/N \lhd G/N$, we find $H/N \leq G/N$ such that $G = w^*(G)H$ and $|H/N|$ divides some power of m. Then $H \cap K = N$ and so

$$w(G) \cap K = w(H) \cap K \leq H \cap K = N.$$

As remarked above, the subgroups like N intersect in $\{1\}$. Hence $w(G) \cap K = 1$ and so $|w(G)| \leq |G/K| < \infty$.

Chapter 2

Verbally elliptic classes

2.1 Virtually nilpotent groups

Theorem 2.1.1 (Romankov) *Every finitely generated virtually nilpotent group is verbally elliptic.*

Proof. Let w be a word in k variables, E a finitely generated group and G a nilpotent normal subgroup of finite index in E. Choose a transversal Y to E/G. Say

$$Y^{(k)} = \{\mathbf{h}_1, \ldots, \mathbf{h}_t\}$$

(where $t = |E : G|^k$), and define generalized word functions $\xi_1, \ldots, \xi_t$ and η on G by

$$\xi_i(\mathbf{g}) = w'_{\mathbf{h}_i}(\mathbf{g}) \quad (\mathbf{g} \in G^{(k)}),$$
$$\eta = \xi_1 \circledast \cdots \circledast \xi_t.$$

Then $G_{\xi_i} \subseteq E_w^{*2}$ for each i and

$$G_\eta \subseteq G_{\xi_1} \cdot \ldots \cdot G_{\xi_t} \cup G_{\xi_t} \cdot \ldots \cdot G_{\xi_1} \subseteq E_w^{*2t}.$$

According to Theorem 1.3.2, η has finite width m, say, in G. Hence

$$\eta(G) = G_\eta^{*m} \subseteq E_w^{*2tm}.$$

With Lemma 1.1.1 this gives

$$w'_E(G) = w'_Y(G) = \langle \xi_1(G), \ldots, \xi_t(G) \rangle$$
$$= \eta(G) \subseteq E_w^{*2tm}.$$

Now write $\overline{} : E \to E/w'_E(G) = \overline{E}$ for the natural map. Then $\overline{G} \leq w^*(\overline{E})$, so $\overline{E}_w$ is finite. But $\overline{E} \in (\mathrm{R}\mathfrak{F})^Q$; it follows by Theorem 1.4.2 that $w(\overline{E})$ has finite order n, say, and then by Lemma 1.1.2 that $w(\overline{E}) = \overline{E}_w^{*n}$.

18

Hence

$$w(E) = E_w^{*n} \cdot w_E'(G)$$
$$\subseteq E_w^{*n} \cdot E_w^{*2tm} = E_w^{*(n+2tm)}.$$

The result follows. ∎

A similar argument gives

Proposition 2.1.2 *Every virtually abelian group is verbally elliptic.*

Proof. Keep the notation of the last proof, assuming now that G is abelian but not assuming that E is finitely generated. Now every generalized word has width 1 in G, so the argument gives

$$w_E'(G) \subseteq E_w^{*2t}.$$

Replacing E by $\overline{E} = E/w_E'(G)$, we may therefore assume that G is marginal for w. Then $E_w = H_w$ is finite and $w(G) = w(H)$ where $H = \langle Y \rangle$ is finitely generated. As before, we may infer that w is concise in H and hence that $w(H)$ is finite. The result then follows as above. ∎

From Theorem 2.1.1 we may deduce the more general

Theorem 2.1.3 *Every virtually nilpotent group of finite rank is verbally elliptic.*

Proof. Let w be any word. Suppose E is a group of rank r and N is a nilpotent normal subgroup of finite index, with nilpotency class c. Let F be the free group on r generators, and put $K = \bigcap \ker \theta$ where θ ranges over the finitely many homomorphisms $F \to E/N$. Then $\gamma_{c+1}(K) \leq \ker \phi$ for every homomorphism $\phi : F \to E$. Now $\widetilde{F} = F/\gamma_{c+1}(K)$ is finitely generated and virtually nilpotent, so w has finite width m, say, in $\widetilde{F}$.

Every element of $w(E)$ lies in $w(E_1)$ for some finitely generated subgroup E_1 of E. As E has rank r, E_1 is a homomorphic image of F, and hence of $\widetilde{F}$. Therefore $w(E_1) = E_{1,w}^{*m} \subseteq E_w^{*m}$. Thus w has width m in E. ∎

Remark. It may have occurred to the reader that similar results for *nilpotent* groups could have been established without using generalized words at all. Is it worth all that trouble to make the small extension to virtually nilpotent groups? This is a moot point, but the techniques will be used in an essential way in later sections. It is certainly not true in general that a word of finite width in a group G necessarily has finite width in a finite extension of G: see Exercise 3.2.2.

2.2   Group ring stuff

Here we collect some module-theoretic results needed in the following section. The augmentation ideal of a group ring $\mathbb{Z}G$ will be denoted I_G.

   The next two theorems, due to P. Hall, are proved in [LR], §§**4.2**, **4.3**.

Theorem 2.2.1 *Let E be a finitely generated group and A an abelian normal subgroup of E such that $E/A = G$ is virtually polycyclic. Then A is finitely generated as a G-module (via the conjugation action of E).*

Theorem 2.2.2 *Let G be a finitely generated virtually nilpotent group and put $R = \mathbb{Z}G$.*

  (i) *R is right and left Noetherian;*

  (ii) *every simple R-module is finite;*

  (iii) *every finitely generated R-module is residually finite;*

  (iv) *if G is nilpotent then I_G has the weak Artin-Rees property.*

To say that an ideal I has *the weak Artin-Rees property* means: if M is a finitely generated R-module and N is a submodule then there exists $n \in \mathbb{N}$ such that

$$MI^n \cap N \le NI.$$

Proposition 2.2.3 *Let G be a finitely generated nilpotent group and M a finitely generated $\mathbb{Z}G$-module. If $MI_G \le N$ for every maximal submodule N of M then $MI_G^n = 0$ for some finite n.*

Proof. Write $I = I_G$. Suppose the claim is false. As M is Noetherian, we may choose a submodule K of M maximal subject to $MI^n \not\le K$ for all n. Replacing M by M/K, we may now assume that every non-zero submodule of M contains MI^n for some n. We may also assume without loss of generality that G acts faithfully on M.

   Now let $1 \ne z \in \mathrm{Z}(G)$. Then [LR], **4.4.2** shows that $M(z-1)^s = 0$ for some s. But $M(z-1) > 0$, so $MI^n \le M(z-1)$ for some n, and it follows that $MI^{ns} = 0$, contradicting the original hypothesis. ∎

Cohomology

   If G is a group and M is a G-module, then

$$H^0(G, M) = \mathrm{C}_M(G); \quad H_0(G, M) = M/MI_G.$$

M is said to be *perfect* if $M = MI_G$, i.e. if $H_0(G, M) = 0$.

   We also need the first and second cohomology groups; for present purposes, the important facts are

- if $H^1(G, M) = 0$ then all complements to M in the semi-direct product $M \rtimes G$ are conjugate;

- if $H^2(G, M) = 0$ then every extension of M by G splits.

Proposition 2.2.4 *Let G be a nilpotent group and M a finite G-module. If either $H_0(G, M) = 0$ or $H^0(G, M) = 0$ then $H^1(G, M) = H^2(G, M) = H_0(G, M) = H^0(G, M) = 0$.*

Proof. This is part of [LR], **10.3.1**. ∎

Corollary 2.2.5 *Let G be a nilpotent group and M a finite G-module. Then $\mathrm{C}_M(G) = 0$ if and only if M is perfect, in which case every submodule of M is perfect.*

2.3 Results of Peter Stroud and Keith George

In his 1966 thesis [S11], Peter Stroud proved that every finitely generated abelian-by-nilpotent group is verbally elliptic. The proof, which is direct and complicated, has never been published. Ten years later, the result was generalized by Keith George in another unpublished thesis [G1], which also contains a more transparent proof of Stroud's theorem. Here I will follow George's approach to obtain a slight further generalization, by combining it with Romankov's result on virtually nilpotent groups.

Throughout, w denotes a word in k variables.

Theorem 2.3.1 *Every finitely generated virtually abelian-by-nilpotent group is verbally elliptic.*

The proof depends on the next three lemmas.

Lemma 2.3.2 [S11] *Let A be an abelian normal subgroup of a group E and write R for the subring of $\mathrm{End}(A)$ generated by the conjugation action of E. Suppose that R is left Noetherian. Then*

$$w'_E(A) \subseteq E_w^{*n}$$

for some finite n.

Proof. For $\mathbf{g} \in E^{(k)}$ and $\mathbf{a} \in A^{(k)}$ we have

$$w'_{\mathbf{g}}(\mathbf{a}) = \prod_j a_{i_j}^{\varepsilon_j v_j(\mathbf{g})} \tag{2.1}$$

$$= \sum_{i=1}^{k} a_i \phi_i(\mathbf{g})$$

where each ε_j is 1 or -1, the v_j are certain fixed words, and $\phi_i(\mathbf{g})$ is the image in R of $\sum_{i_j=i} \varepsilon_j v_j(\mathbf{g})$. Since R is left Noetherian there exist finitely many elements $\mathbf{h}_{ij} \in E^{(k)}$ $(1 \le j \le t, \, 1 \le i \le k)$ such that

$$\sum_{\mathbf{g} \in E^{(k)}} R\phi_i(\mathbf{g}) = \sum_{j=1}^{t} R\phi_i(\mathbf{h}_{ij}).$$

Then

$$w'_E(A) = \sum_{i=1}^{k}\sum_{j=1}^{t} A\phi_i(\mathbf{h}_{ij}),$$

so each element of $w'_E(A)$ takes the form

$$\sum_{i,j} b_{ij}\phi_i(\mathbf{h}_{ij}) = \sum_{i,j} w'_{\mathbf{h}_{ij}}(\underline{b_{ij}}^{(i)})$$

for some $b_{ij} \in A$. Since each value of w' lies in E_w^{*2} the result follows with $n = 2kt$. ∎

Lemma 2.3.3 [S11] *Let H be a finitely generated abelian-by-nilpotent group. Then there exists c such that $\gamma_c(H)$ is abelian and $\gamma_c(H) \cap \mathrm{Z}(H) = 1$.*

Proof. Say $A = \gamma_r(H)$ is abelian. Then A is a finitely generated module for $\mathbb{Z}G$ where $G = H/A$. Put $U = \mathrm{C}_A(G)$. By the weak Artin-Rees property there exists n such that $AI_G^n \cap U \subseteq UI_G = 0$. This translates as $\gamma_{r+n}(H) \cap \mathrm{Z}(H) = 1$. ∎

Lemma 2.3.4 *Let A be a finitely generated module for a finitely generated nilpotent group G. Let $\mathcal{K}$ be the set of G-submodules B of finite index in A such that A/B is perfect, and put $D = \bigcap \mathcal{K}$. Then*

   (i) *$\mathcal{K}$ is closed under finite intersections;*

   (ii) *$DI_G^n = 0$ for some n.*

Proof. Write $I = I_G$. (i) Let $B_1, \ldots, B_r \in \mathcal{K}$ and put $C = B_1 \cap \ldots \cap B_r$. Then A/C is isomorphic to a submodule of the perfect module $A/B_1 \oplus \cdots \oplus A/B_r$. So A/C is perfect by Corollary 2.2.5.

(ii) In view of Proposition 2.2.3, it suffices to show that I kills every simple quotient of D. If $D = 0$ there is nothing to prove; otherwise, replacing A by A/T where T is a maximal submodule of D, we may suppose that D is a simple $\mathbb{Z}G$-module, and have to show that $DI = 0$.

Let N be a submodule of A maximal subject to $N \cap D = 0$. As A/N is residually finite and D is finite, A/N is finite. Also $N \notin \mathcal{K}$ since $N \cap D = 0$, so A/N cannot be perfect. It follows by Corollary 2.2.5 that $C/N := \mathrm{C}_{A/N}(G) > 0$. Therefore $C \ge D$ and so $DI \le N \cap D = 0$ as required. ∎

We are now ready to prove Theorem 2.3.1. Let E be a finitely generated group and A_1 an abelian normal subgroup with E/A_1 virtually nilpotent. The group ring $\mathbb{Z}(E/A_1)$ is left Noetherian, so by Lemma 2.3.2 there exists n such that $w'_E(A_1) \subseteq E_w^{*n}$. Thus to show that w has finite width in E it will suffice to show that it has finite width in $E/w'_E(A_1)$. Thus replacing E by this quotient, we may assume that $w'_E(A_1) = 1$, i.e. that $A_1 \le w^*(E)$.

By Lemma 2.3.3, there exist normal subgroups $A \le H$ of E, with $A \le A_1$, such that E/H is finite, $A = \gamma_c(H)$ and $A \cap Z(H) = 1$. Thus $G = H/A$ is finitely generated nilpotent and A is finitely generated as a G-module. Let $\mathcal{K}$ be the set of G-submodules of A defined in Lemma 2.3.4, and let $\mathcal{K}_0$ denote the subset consisting of E-invariant members of $\mathcal{K}$. Lemma 2.3.4(i) shows that $\mathcal{K}_0 = \{B_0 \mid B \in \mathcal{K}\}$ where $B_0 = \bigcap_{x \in E} B^x$; hence by part (ii) of the lemma we have

$$\bigcap \mathcal{K}_0 = \bigcap \mathcal{K} = 0 \qquad (2.2)$$

(for if $DI^n = 0$ then $DI^{n-1} \le A \cap Z(H) = 0$).

Let $B \in \mathcal{K}_0$. Then $A/B = M$ satisfies $H_0(G, M) = 0$. It follows by Proposition 2.2.4 that H/B splits over A/B and that all complements are conjugate; applying the Frattini argument we infer that E/B splits over A/B, so $E = AL$ and $A \cap L = B$ for some subgroup L. As $A \le w^*(E)$ we then have

$$w(E) \cap A = w(L) \cap A \le L \cap A = B.$$

It now follows by (2.2) that $w(E) \cap A = 1$.

Since E/A is finitely generated and virtually nilpotent, w has finite width l, say, in E/A by Theorem 2.1.1. Hence

$$w(E) \subseteq E_w^{*l}(w(E) \cap A) = E_w^{*l}.$$

This completes the proof.

Keith George extended Stroud's theorem to groups that need not be finitely generated. The key step is the following variation of Lemma 2.3.2:

Lemma 2.3.5 *Let A be an abelian normal subgroup of a group E such that E/A is virtually nilpotent of finite rank. Then there exists n such that*

$$w'_E(A) \subseteq E_w^{*n}.$$

Proof. As we saw in the proof of Theorem 2.1.3, there exists a finitely generated virtually nilpotent group H (previously denoted $\widetilde{F}$) such that every finitely generated subgroup of E/A is a homomorphic image of H. For $\mathbf{h} \in H^{(k)}$ and $1 \le i \le k$ put

$$\psi_i(\mathbf{h}) = \sum_{i_j = i} \varepsilon_j v_j(\mathbf{h}) \in \mathbb{Z}H$$

where the ε_j and v_j are as in (2.1). Since $\mathbb{Z}H$ is left Noetherian, there exist finitely many elements $\mathbf{h}_{ij} \in H^{(k)}$ $(1 \le j \le t, 1 \le i \le k)$ such that

$$\sum_{\mathbf{h} \in H^{(k)}} \mathbb{Z}H \psi_i(\mathbf{h}) = \sum_{j=1}^{t} \mathbb{Z}H \psi_i(\mathbf{h}_{ij}).$$

Now let E_1/A be a finitely generated subgroup of E/A, choose an epimorphism $\pi : H \to E_1/A$, and consider A as an H-module via π. Then for $\mathbf{g} \in E_1^{(k)}$ and $a \in A$ we have

$$a\phi_i(\mathbf{g}) = a\psi_i(\mathbf{h})$$

where $\phi_i(\mathbf{g})$ is the endomorphism defined in the proof of Lemma 2.3.2 and $g_j A = h_j \pi$ $(j = 1, \ldots, k)$. As before we see that each element of $w'_{E_1}(A)$ takes the form

$$\sum_{i,j} b_{ij} \psi_i(\mathbf{h}_{ij}) = \sum_{i,j} b_{ij} \phi_i(\mathbf{g}_{ij})$$

$$= \sum_{i,j} w'_{\mathbf{g}_{ij}} (\underline{b_{ij}}^{(i)})$$

for some $b_{ij} \in A$, where $(g_{ij})_l A = (h_{ij})_l \pi$ $(i, l = 1, \ldots, k, \ j = 1, \ldots, t)$.

Thus as before we see that $w'_{E_1}(A) \subseteq E_{1,w}^{*2kt} \subseteq E_w^{*2kt}$. The result follows since every element of $w'_E(A)$ lies in $w'_{E_1}(A)$ for some subgroup E_1 of E with E_1/A finitely generated. ∎

Theorem 2.3.6 *Let E be an extension of an abelian group by a virtually nilpotent group of finite rank. Then E is verbally elliptic.*

Proof. Let A be an abelian normal subgroup of E such that E/A is virtually nilpotent of rank r. In view of the preceding lemma, to show that w has finite width in E, it will suffice to show that w has finite width in $E/w'_E(A)$; so replacing E by this quotient we may assume that A is marginal for w.

Let N/A be a nilpotent normal subgroup of finite index in E/A, with nilpotency class c, say. Let F be the free group on r generators and define $K \lhd F$ as in the proof of Theorem 2.1.3. Put $\widetilde{F} = F/\gamma_{c+1}(K)'$; then every r-generator subgroup of E is a homomorphic image of $\widetilde{F}$. Also $\widetilde{F}$ is finitely generated and virtually abelian-by-nilpotent, so Theorem 2.3.1 shows that w has finite width m, say, in $\widetilde{F}$.

Now every element of $w(E)$ lies in $w(E_1)$ for some finitely generated subgroup E_1 of E. If E_1 is such a subgroup then AE_1/A is an r-generator group, so $E_1 \le A.\widetilde{F}\pi$ where π is some homomorphism $\widetilde{F} \to E$. As A is marginal for w we then have

$$w(E_1) \subseteq w(\widetilde{F}\pi) = w(\widetilde{F})\pi = (\widetilde{F}_w^{*m})\pi \subseteq E_w^{*m}.$$

The result follows. ∎

Philip Hall proved several theorems about finitely generated soluble groups. In each case, he established that his result was (in a crude way) best possible by exhibiting suitable counterexamples. For example, one of his theorems states that all finitely generated abelian-by-nilpotent groups ($\mathfrak{A}\mathfrak{N}$ groups) are residually finite. He went on to show that 'residual finiteness' is what he called an '$\mathfrak{A}\mathfrak{N}$ property': if $\mathcal{C}$ is a variety of groups defined by 'commutator subgroup laws', then every finitely generated group in $\mathcal{C}$ has the given property if and only if $\mathcal{C} \subseteq \mathfrak{A}\mathfrak{N}$. It is not hard to verify that any such variety either is contained in $\mathfrak{A}\mathfrak{N}$ or else contains all centre-by-metabelian groups (*Exercise!*); so to show that residual finiteness is an $\mathfrak{A}\mathfrak{N}$ property, he only had to establish the existence of one finitely generated centre-by-metabelian group not having the property. (A group G satisfies a 'commutator subgroup law' if $[[\ldots [G, G, \ldots], [G, \ldots] \ldots] \ldots] = 1$ for some fixed bracketing; for example, G is centre-by-metabelian if $[G'', G] = 1$; G is abelian-by-(nilpotent of class $\leq c$) if $[\gamma_{c+1}(G), \gamma_{c+1}(G)] = 1$.)

Similarly, to show that verbal ellipticity is an $\mathfrak{A}\mathfrak{N}$ property it would suffice to solve

Problem 2.3.1 *Find a finitely generated centre-by-metabelian group that is not verbally elliptic.*

It may be that such groups don't exist. We shall see in §3.2 that the two-generator free group in the variety $\mathfrak{N}_2\mathfrak{A}$ (consisting of groups G with $\gamma_3(G') = 1$) is not verbally elliptic. This leaves open the possibility that Stroud's theorem might be generalized in the following way:

Problem 2.3.2 *Prove that every finitely generated group G such that*

$$[\gamma_n(G)', {}_m G] = 1 \tag{2.3}$$

for some $n, m \in \mathbb{N}$ is verbally elliptic.

This would imply that verbal ellipticity is a 'hypercentre-by-$\mathfrak{A}\mathfrak{N}$ property', in view of

Exercise. Show that any commutator subgroup law *either* implies one of the form (2.3) *or* is implied by $\gamma_3(G') = 1$.

2.4 Extracting roots in nilpotent groups

A well-known theory, going back to A. I. Mal'cev, describes the embedding of a torsion-free nilpotent group G in a radicable group $G^{\mathbb{Q}}$, its 'Mal'cev completion' (see for example [LR], §2.1). A consequence of the theory is that for each natural number n, there is a canonical way to embed G in a minimal (torsion-free nilpotent) group $G^{1/n}$ in which each element of G has an nth root (which is unique). For later application, we need a mild generalization of this construction, which allows for some torsion in the group G. (This is included for completeness, as I

could not find a reference; the following theorem and its rather long proof can
be skipped if the reader is willing to restrict attention to virtually torsion-free
groups in §2.6 below.)

Theorem 2.4.1 *Let G be a nilpotent group of class c, and assume that the
torsion subgroup T of G is abelian and divisible. Let $n \in \mathbb{N}$. Then there exist a
nilpotent group $\widetilde{G}$ of class at most c and a homomorphism $\phi : G \to \widetilde{G}$ such that*

(i) *$\ker \phi$ has finite exponent, bounded in terms of n and c;*

(ii) *$T\phi$ is the torsion subgroup of $\widetilde{G}$;*

(iii) *every element of $G\phi$ has an nth root in $\widetilde{G}$;*

(iv) *the elements $h \in \widetilde{G}$ such that $h^n \in G\phi$ generate $\widetilde{G}$;*

(v) *$\widetilde{G}^{n^{c(c+1)/2}} \leq G\phi$;*

(vi) *there is a homomorphism $* : \mathrm{Aut}(G) \to \mathrm{Aut}(\widetilde{G})$ such that $(g\phi)\alpha^* = (g\alpha)\phi$
for each $\alpha \in \mathrm{Aut}(G)$, $g \in G$; if α is the inner automorphism induced by
$h \in G$ then α^* is the inner automorphism induced by $h\phi$.*

I will write $\widetilde{G} = G^{(1/n)}$ and $\phi = \phi_n$.

Remark. If G is torsion-free, then $G^{(1/n)} = G^{1/n}$, the subgroup of $G^{\mathbb{Q}}$
generated by all nth roots of elements of G, and ϕ_n is just the embedding of G in
$G^{1/n}$. In general, ϕ_n induces the embedding of G/T in $G^{(1/n)}/T\phi_n = (G/T)^{1/n}$
(see Exercise 2.4.1).

We will need some elementary combinatorial results.

Proposition 2.4.2 *Let G be a nilpotent group of class at most c, and let $n \in \mathbb{N}$.*

(i) *If $G = \langle X \rangle$ and $H = \langle x^n \mid x \in X \rangle$ then $G^{n^{c(c+1)/2}} \leq H$.*

(ii) *If $x, y \in G$ and $x^n = y^n$ then $(x^{-1}y)^{n^c} = 1$.*

(iii) *$(G')^{n^{2(c-1)}} \leq (G^n)'$.*

Proof. Write $G_i = \gamma_i(G)$. (i) This is clear if $c = 1$. Let $c > 1$ and suppose
inductively that $G^{n^{c(c-1)/2}} \leq HG_c$. Now G_c is central in G, and is generated
by elements $[x_1, \ldots, x_c]$ with each $x_i \in X$, and $[x_1, \ldots, x_c]^{n^c} = [x_1^n, \ldots, x_c^n]$; so
$G_c^{n^c} = \gamma_c(H)$ and

$$G^{n^{c(c+1)/2}} \leq (HG_c)^{n^c} = H^{n^c}\gamma_c(H) \leq H.$$

(ii) We may as well assume that $G = \langle x, y \rangle$. If $c = 1$ the result is clear.
Let $c > 1$ and suppose inductively that $(x^{-1}y)^{n^{c-1}} \in G_c$. Arguing as above,

with $X = \{x, y\}$, we see that $G_c^n = 1$ (it is generated by elements of the form $[x^n, y, x_3 \ldots, x_c]$ with each $x_i \in X$, and $[x^n, y] = [y^n, y] = 1$). The result follows.

(iii) We may as well assume that G^n is abelian. Let $i \geq 2$. Then G_i is generated modulo G_{i+1} by elements of the form $[g_1, \ldots, g_i]$, and $[g_1, \ldots, g_i]^{n^2} \equiv [g_1^n, g_2^n, g_3, \ldots, g_i] = 1 \pmod{G_{i+1}}$. Therefore G_i/G_{i+1} has exponent dividing n^2. The result follows. ∎

Lemma 2.4.3 *Let G be a group and $m \in \mathbb{N}$. Let $I = \mathbb{Z}G(G-1)$ be the augmentation ideal of the group ring $\mathbb{Z}G$. Then for each $c \in \mathbb{N}$ we have*

$$m^{2(c-1)}\mathbb{Z}G \subseteq I^c + \mathbb{Z}G^m.$$

Proof. If $h \in G$ then $m(h-1) \equiv h^m - 1 \pmod{I^2}$; hence

$$mI \subseteq J + I^2$$

where $J = \mathbb{Z}(G^m - 1)$. It follows by induction on n that for $n \geq 1$,

$$m^n I \subseteq J + \sum_{i=0}^{[n/2]} J^i I^{n+1-2i}.$$

Taking $n = 2(c-1)$ and noting that $J \subseteq I$ we deduce that

$$m^{2(c-1)}I \subseteq J + I^c,$$

and the result follows since $\mathbb{Z}G = I + \mathbb{Z}$ and $\mathbb{Z}G^m = J + \mathbb{Z}$. ∎

Corollary 2.4.4 *Suppose that*

$$R' \leq B < R \lhd G$$

where G is a nilpotent group of class c. If $G^m \leq \mathrm{N}_G(B)$ then the normal closure $\overline{B}$ of B satisfies

$$\overline{B}^{m^{2(c-1)}} \leq B.$$

Proof. Assume without loss of generality that $R' = 1$, and consider R as an additively written G-module. Then the lemma gives

$$m^{2(c-1)}\overline{B} = m^{2(c-1)}B\mathbb{Z}G \subseteq B(I^c + \mathbb{Z}G^m) = B$$

since $BI^c = [B, _c G] = 0$. ∎

Proposition 2.4.5 *Let G be a nilpotent group of class at most c, and let $n \in \mathbb{N}$. Let $K \leq G$ and put $\overline{K} = \langle K^G \rangle$. If $\mathrm{N}_G(K) \geq G^n$ then $\overline{K}^{n^f} \leq K$, where $f = f(c)$ depends only on c.*

Proof. Suppose $h = h(c) \in \mathbb{N}_0$ is such that

$$x \in K, \ g \in G \Longrightarrow (x^g)^{n^h} \in K. \tag{2.4}$$

Then Proposition 2.4.2(i) shows that $f(c) = hc(c+1)/2$ will do.

We define $h(c)$ recursively, taking $h(1) = 0$. Let $c > 1$, and put $Z = Z(G)$, $R = \overline{K}Z$. Write $\lambda = n^{h(c-1)}$ and $\mu = n^{f(c-1)}$, chosen inductively so that

$$x \in K, \ g \in G \Longrightarrow (x^g)^{\lambda} \in KZ,$$
$$R^{\mu} \leq KZ.$$

Then Proposition 2.4.2(iii) gives

$$(R')^{\mu^{2(c-1)}} \leq (KZ)' \leq K.$$

Now let $x \in K, \ g \in G$. Then $(x^g)^{\lambda} = az$ with $a \in K$ and $z \in Z \cap \overline{K}$. Taking $B = KR'$ in Corollary 2.4.4 gives

$$\overline{K}^{n^{2(c-1)}} \leq KR'$$

so there exist $b \in K$ and $v \in R'$ such that $z^{n^{2(c-1)}} = bv$. Then

$$\left(z^{n^{2(c-1)}} \right)^{\mu^{2(c-1)}} = b^{\mu^{2(c-1)}} v^{\mu^{2(c-1)}} \in K.$$

Taking

$$h = h(c) = h(c-1) + 2(c-1)(1 + f(c-1))$$

we get $(x^g)^{n^h} = a^{n^h} z^{n^h} \in K$, giving (2.4). ■

Lemma 2.4.6 *Let $A \leq B$ where B is nilpotent of class at most c and A is abelian and divisible. Suppose that $B^e \leq A$ for some $e \in \mathbb{N}$. Let $D = \{x \in B \mid x^e \in B'\}$. Then $B = AD$ and $D^{e^{2c-1}} = 1$.*

Proof. Proposition 2.4.2(iii) shows that B' has exponent dividing $e^{2(c-1)}$, giving the second claim. If $b \in B$ then $b^e = a^e$ for some $a \in A$ since A is divisible; so $(a^{-1}b)^e \in B'$ and so $b \in aD$, which establishes the first claim. ■

Proof of Theorem 2.4.1. Let G be a nilpotent group of class c, and fix $n \in \mathbb{N}$. We start by constructing a 'universal' group generated by nth roots of elements of G, as follows. Let $X = \{x_g \mid 1 \neq g \in G\}$ be a set bijective with $G \smallsetminus \{1\}$ and let F be the free $\mathfrak{N}_c$-group on X (the relatively free group in the variety of nilpotent groups of class at most c). Define the endomorphism $\theta : F \to F$ by $x\theta = x^n$ ($x \in X$). Noting that each lower central factor $\gamma_i(F)/\gamma_{i+1}(F)$ is torsion-free, and θ induces on it the mapping $y \mapsto y^{n^i}$, it is easy to see that $\ker \theta = 1$. Thus

$$H := F\theta = \langle x_g^n \mid 1 \neq g \in G \rangle$$

is (relatively) free on the displayed generating set. Hence there is an epimorphism $\psi : H \to G$ with $x_g^n \psi = g$ for $1 \neq g \in G$. Note for later reference that $F^e \leq H$ where $e = n^{c(c+1)/2}$, by Proposition 2.4.2(i).

Set $K = \ker \psi$ and put $\overline{K} = \langle K^F \rangle$. Proposition 2.4.5 shows that $\overline{K}^m \leq K$ where $m = e^{f(c)}$.

The group $F/\overline{K}$ is a good candidate for the 'universal nth-root group' of G (see Exercise 2.4.1, below); it is not quite good enough because there may be some unwanted extra torsion. Assume now that the torsion subgroup T of G is abelian and divisible. We have $T = S\psi \cong S/K$ where S/K is the torsion subgroup of H/K. Note that $K \leq H \cap \overline{K} \leq S$, so $S\overline{K}/\overline{K} \cong S/(H \cap \overline{K})$ is a divisible abelian group.

Let $R/\overline{K}$ be the torsion subgroup of $F/\overline{K}$. Clearly $R \cap H = S$, and so $R^e \leq S \leq S\overline{K}$. Now apply Lemma 2.4.6 with $R/\overline{K}$ for B and $S\overline{K}/\overline{K}$ for A: taking

$$N = \left\{ x \in R \mid x^e \in R'\overline{K} \right\}$$

we see that $R = SN$ and $N^{e^{2c-1}} \leq \overline{K}$.

Finally, put $\widetilde{G} = F/N$ and define $\phi : G \to \widetilde{G}$ by $g\phi = Nx_g^n$. Then $G\phi = HN/N$ and $\ker \phi = (N \cap H)\psi$ (see Figure 1, below). It follows that $\ker \phi$ has exponent dividing

$$e^{2c-1} \cdot m = n^{\mu(c)}$$

where $\mu(c) = c(c+1)(2c-1+f(c))/2$. The torsion subgroup of $\widetilde{G}$ is

$$R/N = SN/N = T\phi.$$

Evidently every element of $G\phi$ has an nth root in $\widetilde{G}$ and these generate $\widetilde{G}$.

Now let $\alpha \in \text{Aut}(G)$. Then $x_g \mapsto x_{g\alpha}$ defines an automorphism $\widehat{\alpha}$ of F. Clearly $\widehat{\alpha}$ fixes H, and it fixes K because K is generated by the elements $x_g^n x_h^n x_{gh}^{-n}$ and $x_g^n x_{g^{-1}}^n$ $(g, h \in G)$. Therefore $\widehat{\alpha}$ fixes N, so inducing an automorphism α^* on $F/N = \widetilde{G}$. It is evident that $(g\phi)\alpha^* = (g\alpha)\phi$ for each $g \in G$, and that $\alpha \mapsto \alpha^*$ is a homomorphism.

Suppose that α is the inner automorphism of G induced by $b \in G$. Then

$$(x_g\widehat{\alpha})^n = (x_{g^b})^n$$
$$\equiv (x_g^n)^{x_b^n} \pmod{\overline{K}},$$

so

$$\left((x_g\widehat{\alpha})^{-1} x_g^{x_b^n} \right)^{n^c} \in \overline{K}$$

(Proposition 2.4.2(ii)). As $n^c \mid e$ it follows that $x_g\widehat{\alpha} \equiv x_g^{x_b^n} \pmod{N}$. Thus

$$(Nx_g)\alpha^* = (Nx_g)^{b\phi},$$

so α^* is the inner automorphism induced by $b\phi$. This completes the proof.

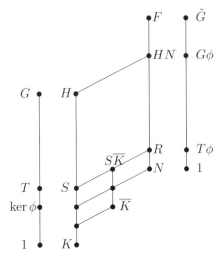

Figure 1

Exercise 2.4.1. Suppose $\xi : G \to L$ is a homomorphism to a nilpotent group L of class at most c such that each element of $G\xi$ has an nth root in L. Show that ξ factors through the homomorphism

$$G \to F/\overline{K}$$
$$g \mapsto x_g^n \overline{K} \quad (1 \neq g \in G).$$

Assume in addition that L is torsion-free and is generated by nth roots of elements of $G\xi$. Put $X = \ker \xi$. Show that there is a unique epimorphism $\pi : \widetilde{G} \to L$ making the following diagram commute:

and that $\ker \pi = Y/N$ where Y is the isolator in F of $X\psi^{-1}$, i.e. the set of $y \in F$ such that $y^t \psi \in X$ for some $t \in \mathbb{N}$. This shows that once X is given, L is essentially unique.

Exercise 2.4.2. Let G be as in Theorem 2.4.1, with upper central series $1 = Z_0 < Z_1 < \ldots < Z_c = G$. Show that $\left| G^{(1/n)} : G\phi \right|$ divides

$$\prod_{i=1}^{c} \left| Z_i : Z_i^{n^{c(c+1)/2}} Z_{i-1} \right|.$$

Deduce that $G\phi$ has finite index in $G^{(1/n)}$ if G has finite rank. [*Hint*: assume without loss of generality that G is torsion-free. Let $\overline{Z}$ denote the centre of $G^{1/n}$, verify that $G^{1/n}\overline{Z}/\overline{Z} \cong (G/Z)^{1/n}$, and argue by induction on c.]

Theorem 2.4.1 will be applied in conjunction with the next result, which shows that a map like ϕ can be extended to an overgroup in which G is normal:

Theorem 2.4.7 *Let* $\phi : G \to \widetilde{G}$ *be a homomorphism of groups such that condition (vi) of Theorem 2.4.1 holds. Suppose that* $G \lhd \Gamma$ *for some group* Γ. *Then* ϕ *extends to a homomorphism* $\phi_* : \Gamma \to \Gamma_1$ *where*

  (i) $\ker \phi_* = \ker \phi \leq G$;

 (ii) $\widetilde{G} \lhd \Gamma_1 = \widetilde{G}.\Gamma\phi_*$;

(iii) $\widetilde{G} \cap \Gamma\phi_* = G\phi$.

Proof. Note first that $\ker \phi$ is characteristic in G, because for $\alpha \in \mathrm{Aut}(G)$ we have

$$g\phi = 1 \implies (g\phi)\alpha^* = 1 \implies (g\alpha)\phi = 1.$$

Hence $\ker \phi \lhd \Gamma$, so replacing G and Γ by $G/\ker \phi$ and $\Gamma/\ker \phi$ we may suppose that ϕ is injective.

Now Γ acts on G by conjugation, and this action extends to an action on $\widetilde{G}$ by

$$\Gamma \to \mathrm{Aut}(G) \xrightarrow{*} \mathrm{Aut}(\widetilde{G}),$$

so that $g \in G \leq \Gamma$ acts on $\widetilde{G}$ as conjugation by $g\phi$. Let $P = \widetilde{G} \rtimes \Gamma$ be the corresponding semi-direct product. Define a map $\lambda : G \to P$ by $g\lambda = (g^{-1}\phi, g)$. One verifies directly that λ is a homomorphism, and that

$$(g\lambda)^\gamma = (g^\gamma)\lambda$$
$$(g\lambda)^h = g\lambda$$

for $g \in G$, $h \in \widetilde{G}$ and $\gamma \in \Gamma$; thus $Q := G\lambda$ is a normal subgroup of P. We put

$$\Gamma_1 = P/Q.$$

Define $\phi_* : \Gamma \to \Gamma_1$ by $\gamma\phi_* = \gamma Q$. Then $\ker \phi_* = \Gamma \cap Q = 1$ since $\ker \phi = 1$. Evidently $\widetilde{G} \cap Q = 1$, so we may identify $\widetilde{G}$ with its image $\widetilde{G}Q/Q \lhd \Gamma_1$. Having made this identification we get

$$\widetilde{G}.\Gamma\phi_* = (\widetilde{G} \cdot \Gamma)/Q = \Gamma_1;$$

while for $g \in G$ we have

$$g\phi_* = (1, g) \cdot Q = g\phi \cdot g\lambda \cdot Q = g\phi \cdot Q = g\phi,$$

so ϕ_* extends ϕ as required. If $\gamma \in \Gamma$ and $\gamma\phi_* \in \widetilde{G}$ then $(1, \gamma) \in \widetilde{G}Q = \widetilde{G} \rtimes G$, so $\gamma \in G$; this establishes (iii). ∎

If G is a torsion-free nilpotent group and ϕ is the embedding of G in $G^{1/n}$ for some $n \in \mathbb{N}$, I will write $G^{1/n}\Gamma$ for the group Γ_1, and consider ϕ_* as an inclusion: thus $G \leq G^{1/n} \leq G^{1/n}\Gamma$ and $\Gamma \leq G^{1/n}\Gamma$. Note that $\Gamma \cap G^{1/n} = G$ in this case.

Exercise 2.4.3. (1) Show that (i)–(iii) determine Γ_1 and $\phi_* : \Gamma \to \Gamma_1$ in Theorem 2.4.7 uniquely up to an isomorphism which is the identity on $\widetilde{G}$. [*Hint*: map P onto Γ_1 in the obvious way.]

(2) Show that $|\Gamma_1 : \Gamma\phi_*| = \left| \widetilde{G} : G\phi \right|$.

(3) Now let $B = B^\Gamma \lhd G$ and suppose that A is a characteristic subgroup of $\widetilde{G}$ with $B\phi \leq A$. Let $\overline{\phi} : \overline{G} = G/B \to \widetilde{G}/A = \left(\widetilde{G} \right)^{-}$ and $\overline{\phi}_* : \overline{\Gamma} = \Gamma/B \to \Gamma_1/A = \overline{\Gamma}_1$ be the homomorphisms induced by ϕ and ϕ_* respectively. Show that the analogues of (i)–(iii) hold with bars everywhere.

2.5 Virtually-nilpotent supplements

An abelian group A has *finite total rank* if the torsion subgroup T of A is a direct sum of finitely many cyclic or quasicyclic groups and A/T has finite rank (so embeds in a finite-dimensional $\mathbb{Q}$-vector space). A soluble group is said to have *finite abelian total rank* (*FATR*) if it has a finite filtration such that each factor is abelian of finite total rank. For soluble groups we have the (proper) implications

$$\text{minimax} \implies \text{FATR} \implies \text{finite rank}$$

(see [LR], Chapter 5). If a soluble FATR group is periodic then it is a *Černikov group*, namely a finite extension of a divisible abelian group with *min* (the minimal condition for subgroups) ([LR], **1.4.1**).

Theorem 2.5.1 *Let G be a group and N a nilpotent FATR normal subgroup such that G/N is virtually nilpotent. Then there exist a group G_1, a nilpotent normal subgroup N_1 of G_1, and a homomorphism $\phi : G \to G_1$ such that $\ker \phi \leq N$, $\ker \phi$ is finite, and*

$$G_1 = N_1 \cdot G\phi = N_1 C$$

where C is a virtually nilpotent subgroup of G_1. Moreover,

$$N\phi \leq N_1 \text{ and } |G_1 : G\phi| = |N_1 : N\phi| < \infty.$$

Corollary 2.5.2 *Suppose that G is virtually a soluble FATR group, with Fitting subgroup N. Then there exist a virtually soluble FATR group G_1 and a homomorphism $\phi : G \to G_1$ with finite kernel such that $G_1 = N_1 \cdot G\phi = N_1 C$ where $N_1 \lhd G_1$ is nilpotent, C is virtually nilpotent, and N_1 contains $N\phi$ as a subgroup of finite index. If G is virtually a minimax group then so is G_1.*

Proof. It is shown in [LR], **5.2.2** that N is nilpotent and G/N is virtually abelian, so the theorem may be applied. Then $G\phi$ is virtually a soluble FATR group (and virtually minimax if G is), and the corollary follows since $|G_1 : G\phi|$ is finite. ∎

At the heart of Theorem 2.5.1 lie some basic splitting and conjugacy results. These are best expressed in cohomological terms. In the next three results, Q denotes a group and M a Q-module that has finite total rank (as an abelian group). For $n \in \mathbb{N}$ we set

$$M[n] = \{a \in M \mid na = 0\}.$$

(When M occurs as a subgroup of another group, I will use additive and multiplicative notation interchangeably for the addition in M.)

Proposition 2.5.3 ([LR], **10.3.6**) *Suppose that Q is nilpotent. If either $H^0(Q, M)$ $(= \mathrm{C}_M(Q))$ or $H_0(Q, M)$ $(= M/[M, Q])$ has finite exponent then so do $H^i(Q, M)$ and $H_i(Q, M)$ for every $i \geq 0$.*

Proposition 2.5.4 *Suppose that $mH^1(Q, M) = 0$, where $m \in \mathbb{N}$. Let $G = M \rtimes Q$ be the semi-direct product, and let H be another complement to M in G.*

(i) *If $mM = M$ then $M[m]H$ is conjugate to $M[m]Q$ in G.*

(ii) *If M is torsion-free then H is conjugate to Q in $M^{1/m}G$.*

Proof. (i) is a special case of [LR], **10.1.10**. (ii) is an exercise. ■

Proposition 2.5.5 *Suppose that $mH^2(Q, M) = 0$, where $m \in \mathbb{N}$. Let G be an extension of M by Q.*

(i) *If $mM = M$ then there exists $H \leq G$ such that*

$$MH = G, \quad M \cap H = M[m].$$

(ii) *If M is torsion-free then $M^{1/m}G$ splits as an extension of $M^{1/m}$ by Q.*

Proof. (i) is a special case of [LR], **10.1.15**. (ii) is an exercise. ■

For brevity, let's say that a normal subgroup N in a group G has an $\mathfrak{NF}$-*supplement* if there exists a virtually nilpotent subgroup H of G such that $NH = G$. We will use the following observation without special mention:

Exercise. If H has a finite normal subgroup K such that H/K is virtually nilpotent then H is virtually nilpotent.

Lemma 2.5.6 *Let $A \triangleleft G$ be groups such that G/A is virtually nilpotent and A is an abelian torsion group of finite total rank. Then A has an $\mathfrak{NF}$-supplement in G.*

Proof. A satisfies the minimal condition on subgroups, so there exists a subgroup B of A minimal subject to $B \triangleleft G$ and G/B virtually nilpotent. Then $B \leq G_1 \triangleleft_f G$ where G_1/B is nilpotent. Also $B^n = B$ for each $n \in \mathbb{N}$: for if $B^n < B$ then B/B^n is finite, so G/B^n is virtually nilpotent; this would contradict the minimal choice of B. Similarly, $B = [B, G_1]$.

Put $Q = G_1/B$. By Proposition 2.5.3 there exists $m_1 \in \mathbb{N}$ such that $m_1 H^i(Q, B) = 0$ for $i = 0, 2$. Similarly, there exists m_2 such that $m_2 H^1(Q, B/B[m_1]) = 0$.

Applying Proposition 2.5.5 we find $H \leq G_1$ with $G_1 = BH$ and $B \cap H = B[m_1]$. Now for any $g \in G$ the subgroup $H^g/B[m_1]$ is a complement to $B/B[m_1]$ in $G_1/B[m_1]$; hence by Proposition 2.5.4 there exists $b \in B$ such that $(HB[m_1 m_2])^g = (HB[m_1 m_2])^b$. It follows that $G = BL$ where

$$L = \mathrm{N}_G(HB[m_1 m_2]).$$

I claim that L is virtually nilpotent; this will complete the proof since clearly $AL = G$.

Let $x \in L \cap B$. Then for $h \in H$ we have

$$[h, x] \in B \cap (HB[m_1 m_2]) = B[m_1 m_2]$$

so $x^{m_1 m_2} \in \mathrm{C}_B(H) = H^0(Q, B)$. Therefore $x^{m_1^2 m_2} = 1$. Thus $L \cap B \leq B[m_1^2 m_2]$; so $L \cap B$ is finite, and the claim follows since $L/(L \cap B) \cong G/B$ is virtually nilpotent. ∎

Lemma 2.5.7 *Let $A \lhd G$ be groups such that G/A is virtually nilpotent and A is torsion-free abelian of finite rank. Then there exists $n \in \mathbb{N}$ such that $A^{1/n}$ has an $\mathfrak{NF}$-supplement in $A^{1/n} G$.*

Proof. Let B be a subgroup of A with minimal rank subject to $B \lhd G$ and G/B virtually nilpotent. Then $B \leq G_1 \lhd_f G$ where G_1/B is nilpotent, and $B/[B, G_1]$ is periodic. Put $Q = G_1/B$, and consider the Q-module $M = \mathbb{Q} \otimes B$. We have $M = [M, Q]$, so applying Proposition 2.5.3 we deduce that $\mathrm{C}_M(Q) = 0$.

It follows that $H^0(Q, B) = \mathrm{C}_B(Q) = 0$; hence, again by Proposition 2.5.3, there exists $m \in \mathbb{N}$ such that $mH^i(Q, B) = 0$ for $i = 1, 2$.

According to Proposition 2.5.5 there exists a complement H for $B^{1/m}$ in $B^{1/m} G_1$. Now $B^{1/m}$ is isomorphic to B as a Q-module, so $mH^1(Q, B^{1/m}) = 0$. Let $g \in G$. Then H^g is another complement for $B^{1/m}$ in $B^{1/m} G_1$; applying Proposition 2.5.4 we find $b \in B^{1/m^2}$ such that

$$H^g = H^b$$

(as subgroups of $B^{1/m^2} G_1$). As in the preceding lemma, this implies that

$$B^{1/m^2} G = B^{1/m^2} L$$

where $L = \mathrm{N}_{B^{1/m^2} G}(H)$.

Arguing as before, we see that

$$L \cap B^{1/m^2} = \mathrm{C}_{B^{1/m^2}}(H) = 1.$$

Thus $L \cong (B^{1/m^2} G)/B^{1/m^2} \cong G/B$, so L is virtually nilpotent. Clearly $A^{1/m^2} L = A^{1/m^2} G$, and the result follows with $n = m^2$. ∎

Lemma 2.5.8 *If T is a periodic nilpotent FATR group then T' is finite.*

Proof. In any nilpotent Černikov group the centre has finite index ([LR], **1.4.4**). It follows that T' is finitely generated, and hence finite since it is a nilpotent torsion group. ∎

Proof of Theorem 2.5.1. Now G is a group and N is a nilpotent FATR normal subgroup of G such that G/N is virtually nilpotent. We have to construct a homomorphism $\phi : G \to G_1$ with certain properties. Suppose that K is a finite normal subgroup of G contained in N and put $\overline{G} = G/K$. If $\overline{\phi} : \overline{G} \to G_1$ has the required properties, then so does $\phi : G \to G_1$ where $g\phi = \overline{g}\overline{\phi}$. Thus we may replace G by G/K. In view of Lemma 2.5.8, we may therefore assume that the torsion subgroup T of N is abelian.

Case 1: where N is torsion-free. In this case, we shall prove that for some $n \in \mathbb{N}$ the normal subgroup $N_1 = N^{1/n}$ has an $\mathfrak{NF}$-supplement C in $G_1 = N^{1/n}G$. The theorem then follows with ϕ the inclusion map $G \to G_1$, in view of Exercise 2.4.3.

If $N = 1$ then G is virtually nilpotent and we may take $n = 1$. Otherwise, let $Z = Z(N)$. Then N/Z is again torsion-free ([LR], **1.2.20**). Arguing by induction on the nilpotency class of N, we may suppose that $(N/Z)^{1/n}$ has an $\mathfrak{NF}$-supplement in $(N/Z)^{1/n}(G/Z)$. Put $N_2 = N^{1/n}$ and $A = Z(N_2)$. Then N_2/A may be identified with $(N/Z)^{1/n}$ and $(N_2G)/A$ may be identified with $(N/Z)^{1/n}(G/Z)$ (see Exercises 2.4.1 and 2.4.3). Thus N_2G has a subgroup L, containing A, such that L/A is virtually nilpotent and $N_2L = N_2G$. Now apply Lemma 2.5.7 with L for G. This shows that for some $m \in \mathbb{N}$, $A^{1/m}$ has an $\mathfrak{NF}$-supplement C, say, in $A^{1/m}L$.

Recall that $N_2^e \le N$ where $e = n^{c(c+1)/2}$ (Theorem 2.4.1(v)). Since A is abelian, this implies that $A^{1/m} \le N^{1/f}$ where $f = me$. Also $N^{1/f} \ge N^{1/n} = N_2$. Hence

$$N^{1/f}C \ge N_2A^{1/m}C = N_2A^{1/m}L \ge G;$$

thus C is an $\mathfrak{NF}$-supplement for $N^{1/f}$ in $N^{1/f}G$.

General case: we are assuming that the torsion subgroup T of N is abelian. By the first case, there exists $n \in \mathbb{N}$ such that $(N/T)^{1/n}$ has an $\mathfrak{NF}$-supplement in $(N/T)^{1/n}(G/T)$. As we saw in the preceding section, there is a homomorphism $\phi = \phi_n : N \to N^{(1/n)} := N_1$ such that $T\phi := A$ is the torsion subgroup of N_1 and N_1/A may be identified with $(N/T)^{1/n}$; moreover, ϕ extends to $\phi_* : G \to G_1 = N_1 \cdot G\phi_*$, where G_1/A may be identified with $(N/T)^{1/n}(G/T)$.

Thus G_1 has a subgroup $L \ge A$ such that $N_1L = G_1$ and L/A is virtually nilpotent. Applying Lemma 2.5.6 with L for G, we find a virtually nilpotent subgroup C of L such that $L = AC$. Then $N_1C = N_1L = G_1$.

Since a nilpotent group of finite rank and finite exponent is finite, both $\ker \phi_* = \ker \phi$ and $|G_1 : G\phi_*| = |N_1 : N\phi|$ are finite (cf. Theorem 2.4.1 and Exercise 2.4.2). Thus writing ϕ for ϕ_* we have established all claims of Theorem 2.5.1.

2.6   Virtually minimax groups

Romankov proved in [R2] that polycyclic groups are verbally elliptic, and Keith
George [G1] did the same for certain classes of soluble minimax groups. Using
a combination of their methods, we can now establish

Theorem 2.6.1 *Let G be a group having a normal subgroup N such that G/N
is virtually abelian and N is a nilpotent minimax group.   Then G is verbally
elliptic.*

Since every soluble minimax group is nilpotent-by-abelian-by-finite ([LR],
5.2.2), we have as a special case

Corollary 2.6.2 *Every virtually soluble minimax group is verbally elliptic.*

The proof depends on

Lemma 2.6.3 *Let $H = \langle S \rangle$ be a nilpotent minimax group.   Then there exists $m \in \mathbb{N}$ such that every finitely generated subgroup of H is contained in
$\langle s_1, \ldots, s_m \rangle$ for some $s_1, \ldots, s_m \in S$.*

Proof. Suppose first that H is abelian and periodic. Then $H = P_1 \times \cdots \times P_q$
where for each i the group P_i is a p_i-group of finite rank r_i, say, and $p_1, \ldots, p_q$
are distinct primes. Let $\pi_i : H \to P_i$ denote the natural projection.

Now let U be a finitely generated subgroup of H. Then $U \leq \langle Y \rangle$ for some
finite subset Y of S. Since $\langle Y \rangle \pi_i$ is a p_i-group, the generating set $Y\pi_i$ contains
a minimal generating set of size at most r_i, so Y contains a subset $Y(i)$ of size
at most r_i such that $\langle Y \rangle \pi_i = \langle Y(i) \rangle \pi_i$. Putting $Z = Y(1) \cup \ldots \cup Y(q)$ we then
have $\langle Z \rangle \pi_i = \langle Y \rangle \pi_i$ for each i. But $|\langle Z \rangle \pi_i|$ is the p_i-part of $|\langle Z \rangle|$ and similarly
for $|\langle Y \rangle \pi_i|$, so $|\langle Z \rangle| = |\langle Y \rangle|$ and hence $\langle Z \rangle = \langle Y \rangle \geq U$. The result follows in
this case with $m = r_1 + \cdots + r_q$, the total rank of H.

Next, assume that H is abelian but not periodic. Let $\{t_1, \ldots, t_r\}$ be a
maximal independent subset of S, so $T = \langle t_1, \ldots, t_r \rangle$ is free abelian of rank r
and H/T is a torsion group, of total rank m_1, say. Let $U \leq \langle Y \rangle$ be as above.
By the first case, $\langle Y \rangle T \leq \langle Z \rangle T$ where Z is a subset of Y of size m_1. Thus

$$U \leq \langle Y \rangle \leq \langle Z \rangle T = \langle Z \cup \{t_1, \ldots, t_r\} \rangle$$

and we have the result with $m = m_1 + r$.

Finally, suppose that H is nilpotent of class $c > 1$. Then $\gamma_c(H)$ is generated
by the set

$$S^\sharp = \langle [s_1, \ldots, s_c] \mid s_1, \ldots, s_c \in S \rangle .$$

Suppose $U \leq \langle Y \rangle$ is as above. Arguing by induction on c, we may suppose that
$\langle Y \rangle \gamma_c(H) \leq \langle Z \rangle \gamma_c(H)$ where Z is a subset of Y of size m_2, a number depending
only on $H/\gamma_c(H)$ and its generating set $\{s\gamma_c(H) \mid s \in S\}$. Then

$$\langle Y \rangle = \langle Z \rangle V$$

where $V = \langle Y \rangle \cap \gamma_c(H)$. Now V is finitely generated because $\langle Y \rangle$ is a finitely generated nilpotent group, and $\gamma_c(H) = \langle S^\sharp \rangle$ is abelian. Hence by the first part we have $V \leq \langle X \rangle$ where X is a subset of $S^\sharp$ of size at most m_3, a number depending only on $\gamma_c(H)$ and $S^\sharp$. Then $X \subseteq \langle Z_1 \rangle$ where $Z_1 \subseteq S$ and $|Z_1| \leq cm_3$; thus

$$U \leq \langle Y \rangle \leq \langle Z \cup Z_1 \rangle,$$

and the result follows with $m = m_2 + cm_3$. ■

(The same result holds if H is merely assumed to be soluble minimax, provided S is a union of conjugacy classes; the proof is similar.)

Lemma 2.6.4 *Let $N \lhd G$ where N is a nilpotent minimax group, and let w be a word. Then*

$$w'_G(N) \subseteq G_w^{*n}$$

for some finite n.

Proof. Say w is a word of length s in k variables, and set

$$S = \left\{ w'_{\mathbf{h}}(\mathbf{a}) \mid \mathbf{h} \in G^{(k)}, \ \mathbf{a} \in N^{(k)} \right\}.$$

Then $\langle S \rangle = w'_G(N) \lhd G$ by Lemma 1.1.1. Let U be a finitely generated subgroup of $w'_G(N)$. According to Lemma 2.6.3, there exist $s_1, \ldots, s_m \in S$ such that $U \leq \langle s_1, \ldots, s_m \rangle$, where m is a number depending only on $w'_G(N)$ and S. Say $s_i = w'_{\mathbf{h}(i)}(\mathbf{a}(i))$, and let ξ_i be the generalized word function on N defined by

$$\xi_i(\mathbf{a}) = w'_{\mathbf{h}(i)}(\mathbf{a}).$$

Put

$$\eta = \xi_1 \divideontimes \cdots \divideontimes \xi_m.$$

Then

$$U \leq \langle \xi_1(\mathbf{a}(1)), \ldots, \xi_m(\mathbf{a}(m)) \rangle \leq \eta(N).$$

Now η is a generalized word function on N of length ms. Hence by Theorem 1.3.2, η has finite width in N, bounded by a number f depending only ms and the rank and class of N. Thus

$$U \subseteq N_\eta^{*f} = \left(\prod_{i=1}^m N_{\xi_i} \cup \prod_{i=m}^1 N_{\xi_i} \right)^{*f} \subseteq G_w^{*2mf};$$

as $U \leq w'_G(N)$ was arbitrary this gives the result with $n = 2mf$. ■

Remark. If N (or just $w'_G(N)$) happens to be a finite p-group, we could take $m = \mathrm{rk}(N)$ in this argument, and deduce in that case that n depends only on s and the rank and class of N.

Corollary 2.6.5 *Let $G = NC$ be a group, where $N \lhd G$, N is a nilpotent minimax group and C is verbally elliptic. Then G is verbally elliptic.*

Proof. If we show that w has finite width l in $G/w'_G(N)$, it will follow that w has width $l+n$ in G where n is given in the lemma. So replacing G by $G/w'_G(N)$ we may now suppose that $w'_G(N) = 1$; that is, that N is marginal for w. Then

$$w(G) = w(C) = C_w^{*l} = G_w^{*l}$$

where $l < \infty$ is the width of w in C.   ∎

Proof of Theorem 2.6.1. We have $N \lhd G$ where N is a nilpotent minimax group and G/N is virtually abelian. Let w be a word and put $W = w(G)$. We will show that w has finite width in G.

Case 1: Suppose first that G is virtually nilpotent. Proposition 2.1.2 shows that G/N is verbally elliptic, so

$$W = (W \cap N) \cdot G_w^{*f}$$

for some finite f. Then $W \cap N$ is generated by the set

$$\begin{aligned} S &= \left(G_w^{*f} \cdot G_w \cdot G_w^{*f} \right) \cap N \\ &= G_w^{*(2f+1)} \cap N. \end{aligned}$$

Now let P be a finitely generated subgroup of $W \cap N$. According to Lemma 2.6.3, there exist $s_1, \ldots, s_q \in S$ such that $P \leq \langle s_1, \ldots, s_q \rangle$, where q depends only on S. As each s_i is a product of $2f+1$ w-values, it follows that $P \leq w(H)$ where $H \leq G$ is generated by $d := kq(2f+1)$ elements (k as usual denoting the number of variables in w).

Say $G_0 \lhd_f G$ is nilpotent of class c. Then F_d has a normal subgroup K of finite index such that $\gamma_{c+1}(K) \leq \ker\phi$ for every homomorphism $\phi : F_d \to G$ (cf. proof of Theorem 2.1.3), and Theorem 2.1.1 shows that w has finite width m, say, in $F_d/\gamma_{c+1}(K)$. Mapping F_d onto H we deduce that w has width m in H. Thus

$$P \leq w(H) = H_w^{*m} \subseteq G_w^{*m}.$$

It follows that $W \cap N \subseteq G_w^{*m}$ and hence that $W = G_w^{*(m+f)}$.

General case: Let $\phi : G \to G_1$ be the homomorphism with finite kernel given by Theorem 2.5.1. In view of Proposition 1.1.3, it will suffice to prove that w has finite width in $G/\ker\phi$; so replacing G by $G\phi$ we may suppose that ϕ is an inclusion map. Thus

$$G \leq G_1 = N_1 G = N_1 C$$

where $N \leq_f N_1 \lhd G_1$, N_1 is nilpotent and C is virtually nilpotent. Then N_1 is a minimax group and

$$\frac{C}{C \cap N_1} \cong \frac{G}{G \cap N_1}$$

is virtually abelian. Case 1 (with C for G) shows that C is verbally elliptic; with Corollary 2.6.5 this implies that w has finite width m, say, in G_1. Let

$$v = w * w^{-1} * w * w^{-1} * \cdots * w * w^{-1}$$

with $2m$ factors (m factors w and m factors w^{-1}). Then

$$G_v \subseteq G_w^{*2m}$$

and

$$v(G) = w(G) \subseteq G \cap G_{1,w}^{*m} \subseteq G \cap G_{1,v}.$$

In particular, w has finite width in G if v does.

There exists $s \in \mathbb{N}$ such that $N \geq N_1^s = M$, say. Lemma 2.6.4 shows that

$$v_G'(M) \subseteq G_v^{*n}$$

for some n. On the other hand, Lemma 1.3.5 shows that

$$v_{\mathbf{g}}'(\mathbf{a})^{s^{c(c+1)/2}} \in v_G'(M) \subseteq v_G'(M)$$

for each $\mathbf{g} \in G^{(k)}$ and $\mathbf{a} \in N_1^{(k)}$, where c is the nilpotency class of N_1. It follows by Proposition 2.4.2(i) that $v_G'(N_1)^e \leq v_G'(M)$ for some $e \in \mathbb{N}$, and so $|v_G'(N_1) : v_G'(M)|$ is finite. Applying Corollary 1.1.4 with $T = v_G'(M)$ and $K = G \cap v_G'(N_1)$, we see that v has finite width in G if it has finite width in G/K.

Let $h \in v(G)$. Then $h \in G_{1,v}$, so $h = v(\mathbf{a}.\mathbf{g}) = v_{\mathbf{g}}'(\mathbf{a})v(\mathbf{g})$ where $\mathbf{a} \in N_1^{(k)}$ and $\mathbf{g} \in G^{(k)}$. But then

$$v_{\mathbf{g}}'(\mathbf{a}) = hv(\mathbf{g})^{-1} \in G \cap v_G'(N_1) = K$$

so $h \equiv v(\mathbf{g}) \bmod K$. Thus v has width 1 in G/K, and the proof is complete.

Chapter 3

Words of infinite width

3.1 Free groups

It is natural to suppose that, apart from silly cases, no word has finite width in every group. This is a fact, but to establish it requires some ingenuity; this was first done by Akbar Rhemtulla in [R1].

What are the 'silly cases'?

Lemma 3.1.1 *Let $w \in F_k$. Then the following are equivalent:*

(a) $G_w = G$ *for every group G;*

(b) $w(G) = G$ *for every group G;*

(c) *there exist integers $e_1, \ldots, e_k$ with $\gcd(e_1, \ldots, e_k) = 1$ such that*

$$w \in x_1^{e_1} \ldots x_k^{e_k} F_k'. \tag{3.1}$$

Proof. Certainly (a) implies (b). Now w satisfies (3.1) for some uniquely determined integers $e_1, \ldots, e_k$, and we may assume that not all e_i are zero (otherwise none of (a), (b), (c) hold). Write $d = \gcd(e_1, \ldots, e_k)$. Then $w(\mathbb{Z}) = \mathbb{Z}_w = d\mathbb{Z}$, so (b) implies (c); and (c) implies (a) since for any $g \in G$ we have

$$g^d \in \langle g \rangle_w \subseteq G_w.$$

■

A word w is *trivial* if w is freely equivalent to 1; let us say that w is *universal* if w satisfies the above conditions (a)–(c), and *silly* if it is either trivial or universal. Silly words have width 1 in every group.

Theorem 3.1.2 *Non-silly words have infinite width in every non-abelian free group.*

It will suffice to show that each non-silly word has infinite width in the free group F of rank 2.

Fix free generators x, y for F and an integer $d > 1$. We define two functions on F as follows.

Let

$$u = \prod_{i=1}^{n} x^{a_i} y^{b_i} \qquad (3.2)$$

where $a_2, \ldots, a_n$ and $b_1, \ldots, b_{n-1}$ are non-zero. Let $i(1) < \ldots < i(s)$ be all the values of $i \in \{1, \ldots, n\}$ with $a_i = 1$; put $\psi_+(u) = s$ (taken to be 0 if $a_i \neq 1$ for every i), and define

$$\psi(u) = \psi_+(u) - \psi_+(u^{-1}).$$

For each $r \in \mathbb{N}$ set

$$\beta_r^+(u) = |\{j < s \mid i(j+1) - i(j) = r\}|,$$
$$\beta_r(u) = \beta_r^+(u) - \beta_r^+(u^{-1})$$

and define

$$\phi(u) = |\{r \mid \beta_r(u) \not\equiv 0 \mod d\}|.$$

Let $u' = \prod_{i=1}^{m} x^{a_i'} y^{b_i'}$ where $a_2', \ldots, a_m'$ and $b_1', \ldots, b_{m-1}'$ are non-zero. If $b_n a_1' \neq 0$ then $u \cdot u'$ is a reduced word, and it is clear that $\psi(u\, u') = \psi(u) + \psi(u')$. More generally, I will say that $u \cdot u'$ is a *non-collapsing product* if $b_n a_1' \neq 0$ or $b_n = a_1' = 0$ or $a_1' = 0 \neq b_n + b_1'$ or $b_n = 0 \neq a_n + a_1'$. Note that if $u \cdot u'$ is non-collapsing then so is $u'^{-1} \cdot u^{-1}$. In this case, it is easy to see that

$$-2 \leq \psi_+(u\, u') - \psi_+(u) - \psi_+(u') \leq 1,$$

and that the number of values of r for which $\beta_r^+(u\, u') \neq \beta_r^+(u) + \beta_r^+(u')$ is at most 3.

Now let g and h be non-identity elements of F. Then we can write $g = g_1 u$ and $h = u^{-1} h_1$ where the three products $g_1 \cdot u, u^{-1} \cdot h_1$ and $g_1 \cdot h_1$ are all non-collapsing. Thus $gh = g_1 h_1$ and

$$\psi_+(g) = \psi_+(g_1) + \psi_+(u) + \varepsilon_1,$$
$$\psi_+(h) = \psi_+(u^{-1}) + \psi_+(h_1) + \varepsilon_2,$$
$$\psi_+(gh) = \psi_+(g_1) + \psi_+(h_1) + \varepsilon_3$$
$$= \psi_+(g) + \psi_+(h) - \psi_+(u) - \psi_+(u^{-1}) + \eta$$

where $-2 \leq \varepsilon_i \leq 1$ for each i and $\eta = \varepsilon_3 - \varepsilon_1 - \varepsilon_2$, so $-4 \leq \eta \leq 5$. Similarly we have

$$\psi_+\left((gh)^{-1}\right) = \psi_+(g^{-1}) + \psi_+(h^{-1}) - \psi_+(u^{-1}) - \psi_+(u) + \eta'$$

where $-4 \leq \eta' \leq 5$. Putting these together gives the first claim in

Lemma 3.1.3 *Let $g, h \in F \smallsetminus \{1\}$. Then*

$$|\psi(gh) - \psi(g) - \psi(h)| \leq 9. \tag{3.3}$$

If c is a product of m commutators in F then

$$|\psi(c)| \leq 9(4m - 1).$$

Proof. For the second claim, write

$$c = g_1^{-1} h_1^{-1} g_1 h_1 g_2^{-1} h_2^{-1} g_2 h_2 \ldots g_m^{-1} h_m^{-1} g_m h_m.$$

Since $\psi(u^{-1}) = -\psi(u)$ for every u, the claim follows from repeated applications of (3.3). ∎

The multiplicative properties of the function ϕ are very similar. Arguing as above, we find that if g and h are non-identity elements of F then

$$\beta_r(gh) = \beta_r(g) + \beta_r(h) \tag{3.4}$$

for all but $2 \times 3 \times 3 = 18$ values of r. As $\beta_r(u^{-1}) = -\beta_r(u)$ for every u, this implies that

$$\beta_r([g, h]) = 0$$

for all but 54 values of r. It is now easy to deduce

Lemma 3.1.4 *Suppose that c is a product of n commutators in F. Then*

$$\beta_r(c) = 0$$

for all but $18(4n - 1)$ values of r. Hence

$$\phi(c) \leq 18(4n - 1).$$

We are now ready to prove the theorem.

Case 1: Suppose first that $w \in F_k$ is a non-trivial commutator word, i.e. that $1 \neq w \in F_k'$. Then w takes a value $g \neq 1$ in $F = F_2$. Below we will prove

Lemma 3.1.5 *Let $1 \neq g \in F$. Then ψ is unbounded on $\langle g^F \rangle$.*

Accepting this for now, we may infer that ψ is unbounded on $w(F)$. On the other hand, since F_w consists of commutators, Lemma 3.1.3 shows that ψ is bounded on F_w^{*m} for each $m \in \mathbb{N}$. It follows that w has infinite width in F.

Proof of Lemma 3.1.5. Let us say that $u \in F$ given by (3.2) is a *standard element* if $n \geq 1$ and $a_1 b_n \neq 0$. If u is standard, then each product $u \cdot u \cdot \ldots \cdot u$ is reduced as written, so for each $m \in \mathbb{N}$ we have $\psi(u^m) = m\psi(u)$. Thus to prove the lemma it will suffice to show that $\langle g^F \rangle$ contains a standard element u with $\psi(u) \neq 0$.

Now the given element $g \neq 1$ has a conjugate g_1 in F which is cyclically reduced. If g_1 is a power of x, then $[g_1, y]$ is standard; if g_1 is a power of y, then $[x, g_1]$ is standard; otherwise, either g_1 is standard or $g_1^{y^c}$ is standard for some c. Thus in any case, $\langle g^F \rangle$ contains a standard element, which we will call u. If $\psi(u) \neq 0$ we are done.

Suppose that $\psi(u) = 0$, and that u is given by (3.2) (so $a_1 \neq 0$). Put $h = u^{x^c} u$ where

$$c = \begin{cases} a_1 - 1 & \text{if} \quad a_1 \neq 1 \\ \\ 2 & \text{if} \quad a_1 = 1 \end{cases}.$$

Then

$$h = \begin{cases} x y^{b_1} \ldots y^{b_n} x^{c+a_1} y^{b_1} \ldots y^{b_n} & \text{if} \quad a_1 \neq 1 \\ \\ x^{-1} y^{b_1} \ldots y^{b_n} x^3 y^{b_1} \ldots y^{b_n} & \text{if} \quad a_1 = 1 \end{cases}.$$

If $a_1 \neq 1$ we see that $\psi(h) = 1$ (if $a_1 \neq -1$) or $\psi(h) = 3$ (if $a_1 = -1$); while if $a_1 = 1$ we have $\psi(h) = -3$. This gives the result, as h is evidently a standard element lying in $\langle g^F \rangle$.

Case 2: Suppose now that $w \in x_1^{e_1} \ldots x_k^{e_k} F_k'$ is a non-commutator word, where $d = \gcd(e_1, \ldots, e_k) \geq 2$. Then every dth power is a w-value, so $w(F)$ contains the element

$$g_n = (xy)^d \prod_{j=2}^n \left((x^j y)^{(j-1)d} (xy)^d \right)$$

for each $n \geq 1$. Now $\beta_r^+(g_n^{-1}) = 0$ for each r, while

$$\beta_r^+(g_n) = \begin{cases} n(d-1) & \text{for} & r = 1 \\ 1 & \text{for} & r = d+1, \ 2d+1, \ (n-1)d+1 \\ 0 & \text{else} \end{cases}.$$

Hence $\phi(g_n) \geq n - 1$ for each n. This shows that ϕ is unbounded on $w(F)$.

Thus to show that w has infinite width in F, it will suffice to establish that ϕ is bounded on each of the sets F_w^{*m}. Now $w = x_1^{e_1} \ldots x_k^{e_k} v_1 \ldots v_q$ where each v_j is a commutator in F_k; so if $\mathbf{u}_i \in F^{(k)}$ $(i = 1, \ldots, m)$ then

$$h := \prod_{i=1}^m w(\mathbf{u}_i) = \prod_{i=1}^m \prod_{j=1}^k u_{ij}^{e_j} \cdot c$$

where c is a product of mq commutators. Using (3.4) repeatedly we see that

$$\beta_r(h) = \sum e_j \beta_r(u_{ij}) + \beta_r(c) \equiv \beta_r(c) \mod d$$

for all but $18m \sum_j |e_j| = f(w, m)$, say, values of r (note that if $e < 0$ then $\beta_r(u^e) = -\beta_r(u^{|e|})$). Thus

$$\phi(h) \leq \phi(c) + f(w, m) \leq 18(4mq - 1) + f(w, m),$$

a bound that depends only on w and m. This completes the proof.

In [R1] Rhemtulla proves more generally that non-silly words have infinite width in every free product $A * B$ apart from $C_2 * C_2$, the dihedral group (this in contrast is verbally elliptic, by Proposition 2.1.2). If one of the free factors contains an element of order greater than 2, the proof is essentially the one I have given above; when both A and B have exponent 2, a more complicated argument is needed.

Exercise. Prove Rhemtulla's theorem in the case where A contains an element x with $x^2 \neq 1$.

3.2   Commutators in p-groups

In a finite group G, a lower bound for the width of a word w is readily obtained by comparing the cardinalities of $w(G)$ and G_w. This might be dignified as the 'Hausdorff dimension method'. If X is a subset of a finite group H, we define the *Hausdorff dimension* of X by

$$\mathrm{h\,dim}_H(X) = \frac{\log|X|}{\log|H|}$$

(with logarithms to any fixed base). The following is self-evident:

Lemma 3.2.1 *If w has width m in a group G and $w(G)$ is finite then* $\mathrm{h\,dim}_{w(G)}(G_w) \geq \frac{1}{m}$.

Proposition 3.2.2 *Let F be the free group on $d \geq 2$ generators and set*

$$G_{d,p} = \begin{cases} F/\gamma_3(F)F^p & \text{(p an odd prime)} \\ \\ F/\gamma_3(F)\gamma_2(F)^2 F^4 & \text{($p = 2$)} \end{cases}.$$

If the word $\gamma_2 = [x, y]$ has width m in $G_{d,p}$ then $m > (d-1)/4$.

(For comparison, we already know that γ_2 has width d in any d-generator nilpotent group; see Corollary 1.2.6.)

Proof. Write $G = G_{d,p}$. Put $\overline{F} = F/\gamma_3(F)$, and set $\mathbf{p} = p$ if p is odd, $\mathbf{p} = 4$ if $p = 2$. Then

$$\overline{F}^{\mathbf{p}} \cap \overline{F}' \leq (\overline{F}')^p$$

(*exercise!*). As $\overline{F}'$ is free abelian of rank $d(d-1)/2$ it follows that G' is elementary abelian of the same rank, while $G/Z(G)$ is elementary abelian of rank d.

It follows that γ_2 takes at most

$$1 + \frac{(p^d - 1)(p^d - p)}{2(p-1)} < p^{2d-1}$$

values in G (note that $[x, y^n] = [x^n, y]$ and $[y, x] = [x, y^{-1}]$ in G); while $|\gamma_2(G)| = p^{d(d-1)/2}$. Hence

$$h \dim_{G'}(G_{\gamma_2}) < \frac{2(2d-1)}{d(d-1)} < \frac{4}{d-1}.$$

The preceding lemma now gives the result. ∎

A sharper result can be obtained by linear algebra, without using finiteness. The following exercise implies a lower bound of $[d/2]$ for the width of γ_2 in $G_{d,p}$, and also in some torsion-free nilpotent groups, such as $F_d/\gamma_3(F_d)$ and its Mal'cev completion.

Exercise 3.2.1. Let $d \geq 2$ and let V be a d-dimensional vector space over a field of characteristic $\neq 2$. Show that the exterior square $\wedge^2 V$ contains elements that cannot be written as a sum of fewer than $[d/2]$ wedges $u \wedge v$ ($u, v \in V$). [Hint: let $\{e_1, \ldots, e_d\}$ be a basis for V. Then $\wedge^2 V$ has basis $f_{ij} = e_i \wedge e_j$ ($1 \leq i < j \leq d$). Show that the f_{ij}-component of

$$\sum_{s=1}^{m} \left(\left(\sum_{i=1}^{d} \lambda_{si} e_i \right) \wedge \left(\sum_{j=1}^{d} \mu_{sj} e_j \right) \right)$$

is the i, j-entry of a certain skew-symmetric matrix that has rank at most $2m$.]

The free group $F = F_d$ has an automorphism of order d which permutes (a chosen set of) free generators cyclically. This automorphism induces an automorphism t on $G_{d,p}$, and we take

$$H_{d,p} = G_{d,p} \rtimes \langle t \rangle$$

to be the corresponding semi-direct product. Obviously $H = H_{d,p}$ is a two-generator group, and H' is nilpotent of class 2. If d is a power of p then H is a p-group.

Proposition 3.2.3 *If the word* $\delta_2 = [[x_1, x_2], [x_3, x_4]]$ *has width* m *in* $H_{d,p}$ *then* $m > (d-2)/4$.

Proof. Write $G = G_{d,p}$ and $Z = Z(G)$. It is easy to see that $H'Z/Z$ is a codimension-one subspace in the $\mathbb{F}_p$-vector space G/Z, and hence that (i) the word γ_2 takes at most $p^{2(d-1)-1}$ values in H', and (ii) $|H''| = p^{(d-1)(d-2)/2}$.

As $H_{\delta_2} \subseteq H'_{\gamma_2}$ we deduce as above that $h \dim_{H''}(H_{\delta_2}) < 4/(d-2)$, and the result follows by Lemma 3.2.1. (We are using the general – and easy – fact that $\delta_2(H) = H''$.) ∎

If Q is any group having $H_{d,p}$ as a quotient, then δ_2 cannot have width $\leq (d-2)/4$ in Q. If p is odd, the same holds for the word

$$\delta_2 * \gamma_2^p = [[x_1, x_2], [x_3, x_4]][x_5, x_6]^p,$$

since γ_2^p vanishes identically on $H_{d,p}$. We may infer

Corollary 3.2.4 *The words δ_2 and $\delta_2 * \gamma_2^p$ have infinite width in the two-generator free $\mathfrak{N}_2\mathfrak{A}$ group and in the two-generator free $\mathfrak{N}_2\mathfrak{A}$ pro-p group (assuming p odd for the second word).*

This was essentially obtained by Stroud in his unpublished thesis [S11], and independently by Romankov in [R2] (he gives an elegant construction using 3×3 matrix groups over finite rings).

Andrei Jaikin realized that the idea of Proposition 3.2.3 can be pushed much further. Using the 'Hausdorff dimension method', he showed in [J1] that *every word $w \in F''(F')^p \setminus \{1\}$ has infinite width in the two-generator free pro-p group; this will be explained in §4.5, once we have discussed the ramifications of verbal width in profinite groups.

The following example, taken from [G1], shows that finite width of a word need not be preserved on making a finite group extension:

Exercise 3.2.2. Let $H = F/\gamma_3(F)F^7$ where F is free on the countably infinite set X. Define $\sigma \in \mathrm{Aut}(H)$ by $x^\sigma = x^2$ $(x \in X)$ and put $G = H \rtimes \langle \sigma \rangle$. Show that the word δ_2 has width 1 in H but infinite width in G, while $|G : H| = 3$. [*Hint*: show that $G' = H$.]

Chapter 4

Words and profinite groups

4.1   Verbal subgroups in profinite groups

The history of group theory shows that finiteness imposes sometimes astonishing restrictions on the algebraic structure of a group – the ultimate manifestation of this being the classification of the finite simple groups. Among the algebraic consequences of finiteness we may therefore expect to find some that relate to the behaviour of word-values.

In a finite group G, every word has width trivially bounded by $|G|$ (Lemma 1.2). The interest therefore shifts to finding bounds, independent of the group order, that hold uniformly over suitable infinite classes of finite groups (also interesting is the *distribution* of word-values in a given finite group; I will not discuss this here, but see [A] and [NS3]). The natural setting for questions of this kind is the theory of *profinite groups*, where uniform properties of a family of finite groups are reflected in topological properties of its inverse limit, an infinite compact group. (For undefined concepts and unproved assertions in the rest of this section, see [DDMS], Chapter 1. For more background, see also [W4], Chapters 0, 1 and [RZ], Chapters 1, 2, 4.)

For a profinite group G we write

$$\mathfrak{F}(G) = \{G/N \mid N \triangleleft_o G\}$$

where $N \triangleleft_o G$ means 'N is an open normal subgroup of G'; thus $\mathfrak{F}(G)$ is the set of all finite continuous quotients of G, and G is the inverse limit of the system $\mathfrak{F}(G)$ with its natural projection maps $G/M \to G/N$ ($M \leq N$). The minimal cardinality of a topological generating set for G is denoted $\mathrm{d}(G)$. A basic illustration of the maxim stated near the end of the previous paragraph is the formula

$$\mathrm{d}(G) = \sup \{\mathrm{d}(Q) \mid Q \in \mathfrak{F}(G)\}; \tag{4.1}$$

this expresses the principle 'G is (topologically) finitely generated if and only if the generating numbers $\mathrm{d}(Q)$ are uniformly bounded as Q ranges over $\mathfrak{F}(G)$'.

(Henceforth, I will say that a profinite group is *finitely generated* if it is *topologically* finitely generated – an infinite profinite group can never be finitely generated as an abstract group, as it is necessarily uncountable.)

Let us explore the topological significance of finite width for a word w. For any group H, let

$$m_w(H)$$

denote *the* width of w in H, i.e. the minimal $m \in \mathbb{N}$ such that w has width m, or ∞ if there is no such m.

Lemma 4.1.1 *Let G be a profinite group. Then for each $n \in \mathbb{N}$ the set G_w^{*n} is closed in G.*

Proof. For each n-tuple $(\varepsilon_1, \ldots, \varepsilon_n) \in \{\pm 1\}^{(n)}$ we have a continuous mapping

$$w_\varepsilon : G^{(nk)} \to G$$
$$(\mathbf{g}_1, \ldots, \mathbf{g}_n) \mapsto w(\mathbf{g}_1)^{\varepsilon_1} \ldots w(\mathbf{g}_n)^{\varepsilon_n}.$$

Since $G^{(nk)}$ is compact, the image X_ε of w_ε is closed in G; hence so is

$$G_w^{*n} = \bigcup_{\varepsilon \in \{\pm 1\}^{(n)}} X_\varepsilon.$$

∎

Now suppose that G is a profinite group and that

$$m_w(Q) \le m \text{ for all } Q \in \mathfrak{F}(G).$$

Then $NG_w^{*(m+1)} = NG_w^{*m}$ for every $N \lhd_o G$. Since G_w^{*m} is closed in G we obtain

$$G_w^{*m} = \bigcap NG_w^{*m} = \bigcap NG_w^{*(m+1)} \supseteq G_w^{*(m+1)}$$

(taking intersections over all $N \lhd_o G$). It follows that $w(G) = G_w^{*m}$. Thus (for a profinite group G)

$$m_w(G) = \sup \{m_w(Q) \mid Q \in \mathfrak{F}(G)\}, \tag{4.2}$$

in perfect analogy with (4.1).

What does this mean topologically? If w has width n then $w(G) = G_w^{*n}$ is closed by Lemma 4.1.1. Suppose, conversely, that $W = w(G)$ is closed. Then the compact Hausdorff space

$$W = \bigcup_{n=1}^{\infty} G_w^{*n}$$

is the ascending union of its closed subspaces G_w^{*n}; it follows by a version of the Baire Category Theorem (cf. [DDMS], Exercise 3.6) that for some n the set G_w^{*n} contains a non-empty open subset U of W. Then U contains hV for some $h \in W$ and some $V \lhd_o W$. In particular, V has finite index in W, so $W = YV$

for some finite set Y. Now $Y \subseteq G_w^{*s}$ and $h \in G_w^{*t}$ for some finite s and t, and we conclude that

$$W \subseteq Yh^{-1}U \subseteq G_w^{*(s+t+n)}.$$

Thus w has width $s + t + n$ in G. So we have established

Proposition 4.1.2 [H1] *The following three conditions are equivalent, for a word w and profinite group G:*

(a) *there is a finite upper bound for $m_w(Q)$, $Q \in \mathfrak{F}(G)$;*

(b) *w has finite width in G;*

(c) *$w(G)$ is closed in G.*

Let $\mathcal{C}$ be a family of finite groups, closed under forming quotients and finite subdirect products (a 'formation'). We say that w is *uniformly elliptic in $\mathcal{C}$* if for each natural number d there is a natural number $f_w(d)$ such that

$$H \in \mathcal{C}, \ \mathrm{d}(H) \leq d \Longrightarrow m_w(H) \leq f_w(d).$$

A *pro-$\mathcal{C}$ group* is a projective limit of groups in $\mathcal{C}$ (the inverse limit of an inverse system with all maps surjective), or equivalently it is a profinite group G such that $\mathfrak{F}(G) \subseteq \mathcal{C}$. (When $\mathcal{C}$ is the class of finite p-groups for some prime p, or the class of finite nilpotent groups, or the class of finite soluble groups, a pro-$\mathcal{C}$ group is said to be *pro-p, pronilpotent,* or *prosoluble,* respectively.)

The *free pro-$\mathcal{C}$ group* on d generators is the pro-$\mathcal{C}$ completion of F_d, namely the inverse limit

$$\widehat{(F_d)}_{\mathcal{C}} = \varprojlim (F_d/N)$$

where F_d/N ranges over all $\mathcal{C}$-quotients of F_d. Every d-generator $\mathcal{C}$-group arises as one of these quotients, so Proposition 4.1.2 yields

Proposition 4.1.3 *Let $\mathcal{C}$ be a formation of finite groups and let w be a word. Then w is uniformly elliptic in $\mathcal{C}$ if and only if the verbal subgroup $w(G)$ is closed in G for every finitely generated pro-$\mathcal{C}$ group G.*

(For more on free pro-$\mathcal{C}$ groups see [W4], Chapter 5 and [RZ], Chapter 3.)

The topological point of view suggests a natural generalization. Any set W of words defines a *group variety* $\mathcal{V}_W$, the class of all groups G such that $w(G) = 1$ for every $w \in W$. For any family of groups $\mathfrak{X}$, let $W_{\mathfrak{X}}$ denote the set of all words w such that $w(G) = 1$ for every $G \in \mathfrak{X}$; then $\mathcal{V}_{W_{\mathfrak{X}}}$ is the *variety generated by $\mathfrak{X}$*. For any variety $\mathcal{V}$ and any group G, we write

$$\mathcal{V}(G) = W_{\mathcal{V}}(G) = \langle w(G) \mid w \in W_{\mathcal{V}} \rangle;$$

this is the smallest normal subgroup N of G such that $G/N \in \mathcal{V}$. For any $d \in \mathbb{N} \cup \{\infty\}$ the *free d-generator $\mathcal{V}$-group* is then $F_d/\mathcal{V}(F_d)$. It is not hard to see that the free $\mathcal{V}$-group on $\aleph_0$ generators generates $\mathcal{V}$ ([N], Theorem 15.62); see §4.4.

If $\mathcal{V}$ is *finitely based*, i.e. defined by a *finite* set of words W, then it can be defined by a single word; indeed, if $W = \{w_1, \ldots, w_n\}$ then $\mathcal{V}_W = \mathcal{V}_{\{u\}}$ where $u = w_1 * w_2 * \cdots * w_n$. In this case, $\mathcal{V}(G) = u(G)$.

The analogue of Proposition 4.1.3 is

Proposition 4.1.4 *Let $\mathcal{C}$ be a formation of finite groups and let $\mathcal{V}$ be a variety. Let $d \in \mathbb{N}$. Then the following are equivalent:*

(a) *the subgroup $\mathcal{V}(G)$ is closed in G for every d-generator pro-$\mathcal{C}$ group G;*

(b) *there exists a word $w \in W_{\mathcal{V}}$ such that $\mathcal{V}(G) = w(G)$ and $\mathcal{V}(G)$ is closed in G for every d-generator pro-$\mathcal{C}$ group G;*

(c) *there exist a word $w \in W_{\mathcal{V}}$ and a natural number m such that $\mathcal{V}(G) = w(G) = G_w^{*m}$ for every d-generator $\mathcal{C}$-group G.*

Proof. Let G be the free pro-$\mathcal{C}$ group on d generators, and suppose that $\mathcal{V}(G)$ is closed in G. For each natural number n put

$$X_n = (G_{w_1} \ldots G_{w_n})^{*n}$$

where $W_{\mathcal{V}} = \{w_n \mid n \in \mathbb{N}\}$. Then

$$\mathcal{V}(G) = \bigcup_{n=1}^{\infty} X_n,$$

an ascending union of closed sets. As before, the Baire Category Theorem ensures that for some n, the set X_n contains a non-empty open subset of $\mathcal{V}(G)$, and hence contains a coset of some open normal subgroup of $\mathcal{V}(G)$. It follows as above that $\mathcal{V}(G) = X_m$ for some $m \geq n$. Putting $w = w_1 * w_1^{-1} * w_2 * w_2^{-1} * \cdots * w_m * w_m^{-1}$ we have

$$\mathcal{V}(G) = X_m \subseteq G_w^{*m} \subseteq w(G) \leq \mathcal{V}(G).$$

As every d-generator $\mathcal{C}$-group is an image of G, this shows that (a) implies (c).

Now assume (c), and let G be an arbitrary d-generator pro-$\mathcal{C}$ group. Then

$$\mathcal{V}(G)N/N = \mathcal{V}(G/N) = G_w^{*m} N/N$$

for every $N \triangleleft_o G$, and using Lemma 4.1.1 we deduce that

$$\mathcal{V}(G) \subseteq G_w^{*m} \subseteq w(G) \leq \mathcal{V}(G).$$

Thus (b) follows; and (b) trivially implies (a). ∎

Of course, some of the ellipticity results we have seen in earlier sections have profinite implications: if a word w has width m in a group G then w has width m in every finite quotient of G. Combined with (4.2) and Proposition 4.1.2, this simple observation gives

Theorem 4.1.5 (i) *Let G be a d-generator pronilpotent group and let $t \geq 2$. Then the word γ_t has width d^{t-1} in G, and the subgroup $\gamma_t(G)$ is closed in G.*

(ii) *Let G be a finitely generated profinite group. If G is virtually abelian-by-nilpotent then G is verbally elliptic, and every verbal subgroup $w(G)$ is closed.*

(The hypothesis in (ii) means that G has an *open* normal subgroup which is abelian-by-nilpotent; but see the exercise below.) Part (ii) is stated for completeness; we will only need the special case where G is virtually nilpotent, which does not depend on the material of §§2.2, 2.3.

Proof. (i) The word γ_t has width d^{t-1} in every d-generator nilpotent group (Corollary 1.2.8).

(ii) Let w be a word, and suppose that the d-generator profinite group G has an open normal subgroup H with $\gamma_{c+1}(H)' = 1$. Put

$$K = \bigcap \ker \theta \lhd F_d$$

where θ ranges over the finitely many epimorphisms from F_d onto the finite group G/H. Theorem 2.3.1 shows that w has finite width m, say, in the virtually abelian-by-nilpotent group $P = F_d/\gamma_{c+1}(K)'$. Now let $N \lhd_o G$. Then there is an epimorphism $\phi : F_d \to G/(N \cap H)$, and $K\phi \leq H/(N \cap H)$; as $\gamma_{c+1}(H)' = 1$ it follows that $\gamma_{c+1}(K)' \leq \ker \phi$. Thus ϕ induces an epimorphism $P \to G/(N \cap H) \to G/N$; it follows that w has width m in G/N. ∎

Corollary 4.1.6 *Let G be a finitely generated pronilpotent group and let w be a word. If G has an open normal subgroup H such that $Hw(G)/w(G)$ is nilpotent then $w(G)$ is closed in G.*

Proof. We have $\gamma_t(H) \leq w(G)$ for some t. The group H is again a finitely generated pronilpotent group, so $\gamma_t(H)$ is closed in H by part (i) of the theorem. Then $\gamma_t(H)$ is closed in G, so $\widetilde{G} = G/\gamma_t(H)$ is a finitely generated, virtually nilpotent, profinite group. Therefore $w(\widetilde{G})$ is closed in $\widetilde{G}$ by part (ii), and the result follows since $w(\widetilde{G}) = w(G)/\gamma_t(H)$. ∎

Exercise. Let $\mathcal{V}$ be a group variety, G a profinite group and H a normal subgroup of G, with closure $\overline{H}$. Show that if $H \in \mathcal{V}$ then $\overline{H} \in \mathcal{V}$. Hence show that if G has a subgroup of finite index that belongs to $\mathcal{V}$ then G has an open normal subgroup belonging to $\mathcal{V}$. [*Hint:* note that if $N \lhd_o G$ then $\mathcal{V}(HN) \leq N$.]

Remark. As we shall see in the next section, the fussy distinction between subgroups of finite index and open subgroups in a finitely generated profinite group will turn out to be unnecessary; but this is a fact that lies much deeper than present considerations. For similar reasons, Corollary 4.1.6 remains true if 'pronilpotent' is replaced by 'profinite'; this will be discussed in §4.7.

The question '*which words are uniformly elliptic in all finite groups?*' is still open; in the following sections we will discuss some partial answers, and some related questions.

Notation

For any subset S of a profinite group G, we will write $\overline{S}$ to denote the closure of S in G. Thus

$$\overline{S} = \bigcap_{N \triangleleft_o G} SN.$$

This is a subgroup (normal subgroup) if S is one.

4.2   Open subgroups

My interest in questions of verbal width was initially motivated by a quite different problem; this might be called the 'rigidity' of profinite groups. To what extent is the topology of a profinite group controlled by the algebraic structure?

Let $C = \langle c \rangle$ be a cyclic group of prime order p and let X be a countably infinite set. The Cartesian power

$$G = C^X$$

is a profinite group, with the product topology (where C has the discrete topology). As an abstract group, G is elementary abelian of uncountable rank $\mathfrak{c}$, hence there are $2^{\mathfrak{c}}$ homomorphisms $G \to C$ (given any subset U of a basis T, we can map each element of U to c and each element of $T \setminus U$ to 1). But G has only countably many open subgroups, since each open subgroup contains a basic open subgroup of the form

$$C^{X \setminus Y} \times \prod_Y 1$$

for some finite subset Y of X, and each such basic open subgroup has finite index. It follows that nearly all the homomorphisms $G \to C$ are not continuous.

The group G is a pro-p group. Around 1975, J-P. Serre showed that such pathology cannot occur in a pro-p group if it is *finitely generated*: indeed in such a group, *every subgroup of finite index is open* (see [S9] §4.2, Exercise 6).

Exercise 4.2.1. Show that the following are equivalent for a profinite group G:

(a) every subgroup of finite index in G is open;

(b) every normal subgroup of finite index in G is open;

(c) every group homomorphism from G to any profinite group is continuous.

A profinite group G satisfying (a)–(c) is said to be *strongly complete* (because then G is its own profinite completion).

Now let w be the word $[x,y]z^p$. Then for any group G the quotient $G/w(G)$ is an elementary abelian p-group, and hence residually finite. So if G happens to be a strongly complete profinite group, then $w(G)$ is the intersection of a family of *open* subgroups; as open subgroups are closed it follows that $w(G)$ is *closed* in G. Thus (in view of Proposition 4.1.3) Serre's theorem implies that w *is uniformly elliptic in the class of finite p-groups.*

The converse is also true; it depends on

Exercise 4.2.2. Let G be a pro-p group and N a normal subgroup of finite index in G. Show that G/N is a p-group. [*Hint*: consider the possible values for the index $|C : C \cap N|$ when C is a procyclic subgroup of G.]

Now we can deduce Serre's theorem, assuming that $w = [x,y]z^p$ is uniformly elliptic in the class of finite p-groups. Let G be a d-generator pro-p group; then $w(G) = W$ is closed in G. If $W \leq T \triangleleft_o G$ then G/T is a d-generator elementary abelian p-group, so $|G/T| \leq p^d$. Hence there is a unique minimal such subgroup T_0, and as W is closed we have $W = T_0$. Thus W is *open* in G; hence W is again a finitely generated pro-p group. Now let N be a proper normal subgroup of finite index in G. Then $G/N = Q$ is a p-group by Exercise 4.2.2, so

$$WN/N = Q'Q^p < Q = G/N.$$

Hence $|W : W \cap N| = |WN/N| < |G : N|$; arguing by induction on the index we may suppose that $W \cap N$ is open in W. It follows that $W \cap N$ is open in G, and then so is N.

Exercise 4.2.3. Prove that the word $[x,y]z^p$ has width $m + 1$ in every m-generator nilpotent group, and hence is uniformly elliptic in the class of nilpotent groups. [Of course, the second statement is a special case of Corollary 4.1.6.]

Serre (loc. cit) raised the question: *is every finitely generated profinite group strongly complete?* To answer it, we need a way of producing closed subgroups that doesn't use the topology; the only known strategy is to use Proposition 4.1.2. Potentially, this provides a collection of closed verbal subgroups $w(G)$ in a profinite group G; but we need more: (1) $w(G)$ should be *open*, not just closed, and (2) given $N \triangleleft_f G$ it should be possible to choose w so that $w(G/N) < G/N$. This motivates the following

Definition The word w is *d-locally finite*, for a natural number d, if $F_d/w(F_d)$ is finite.

Thus w is d-locally finite if every d-generator group in the variety defined by w is finite.

Exercise 4.2.4. Let w be a d-locally finite word and G a d-generator profinite group. Show that the closure $\overline{w(G)}$ of $w(G)$ in G is open.

Exercise 4.2.5. Let Q be a finite group and let $d \in \mathbb{N}$. Construct a d-locally finite word w such that $w(Q) = 1$. [*Hint:* let K be the intersection of the kernels of all homomorphisms from $F_d = \langle x_1, \ldots, x_d \rangle$ into Q. Let $v_1, \ldots, v_n$ be generators for K, and take $w = v_1 * \cdots * v_n$.]

Remark. From this exercise one may deduce that the variety $\mathcal{V}_d(Q)$ defined by all d-variable laws of the finite group Q is finitely based. A much deeper theorem, due to Oates and Powell, shows that in fact the variety generated by Q – namely the intersection of the $\mathcal{V}_d(Q)$ over all $d \in \mathbb{N}$ – is also finitely based; see [N], Chapter 5.

The proof of the following theorem will be discussed in §4.7:

Theorem 4.2.1 [NS1] *Let $d \in \mathbb{N}$ and let w be a d-locally finite word. Then there exists $f = f(w, d) \in \mathbb{N}$ such that w has width f in every d-generator finite group.*

Given this result, it is now easy to deduce

Theorem 4.2.2 *Every finitely generated profinite group is strongly complete.*

Proof. Let G be a d-generator profinite group and N a normal subgroup of finite index. Exercise 4.2.5 provides a d-locally finite word w such that $w(G/N) = 1$. Then $w(G) \leq N$. Now Theorem 4.2.1, with Proposition 4.1.2, shows that $w(G)$ is closed; hence $w(G)$ is open, by Exercise 4.2.4. Therefore N is open in G, and the result follows. ∎

If the profinite group G is strongly complete then its topology is determined by its group structure, in the sense that there is exactly *one* topology on the underlying abstract group G making it a profinite group, namely the given topology. Indeed, condition (c) in Exercise 4.2.1 shows that the identity map $G \to G$ is continuous (relative to the original topology on the domain and an arbitrary topology on the range), and every bijective continuous homomorphism of profinite groups is a topological isomorphism. The following theorem of Jarden and Lubotzky shows that the structure of a finitely generated profinite group is even more rigidly determined, namely by its *elementary theory*. Two (abstract) groups are said to be *elementarily equivalent* if they satisfy the same sentences in the first-order language $\mathcal{L}$ of group theory (first-order logic with symbols $=$, $\cdot$ (for group multiplication), $^{-1}$ (inversion) and 1 (identity element)).

Theorem 4.2.3 [JL] *Let G and H be elementarily equivalent profinite groups. If one of them is finitely generated then G and H are isomorphic as profinite groups.*

This is a surprising result, in stark contrast with the situation in abstract groups, whose elementary theory is usually much too weak to determine the isomorphism type; for example, all finitely generated non-abelian free groups are elementarily equivalent [S7] – while Theorem 4.2.3 shows that they are pairwise distinguished by the elementary theories of their profinite completions.

Let G and H be as above and suppose that $\mathrm{d}(G) = d$ is finite. Let w be a d-locally finite word. As in the proof of Theorem 4.2.2, we see that $w(G)$ is open in G. Taking $f = f(w, d)$ as in Theorem 4.2.1 and using (4.2) we have $w(G) = G_w^{*f}$. Now it is easy to see that the predicate

$$x \in G_w^{*f}$$

can be expressed by a formula $L_{w,f}(x)$ in the language $\mathcal{L}$, and hence so can the statement '$w(G) = G_w^{*f}$', which is equivalent to

$$(\forall x \in G)\ (L_{w,f+1}(x) \implies L_{w,f}(x)).$$

It follows that $w(H) = H_w^{*f}$.

Next, to each finite group F we can associate a sentence $I(F)$ of $\mathcal{L}$ which has the property: if Q is any group then $I(F)$ is true in Q if and only if $Q \cong F$ (*Exercise*: $I(F)$ says that there exist $|F|$ distinct elements which exhaust the group and multiply according to the multiplication table of F). To each sentence S we associate a sentence $S'_{f,w}$ obtained from S by replacing each term of the form $x = y$ by one of the form

$$\exists z (L_{w,f}(z) \wedge x = yz).$$

Thus if K is a group such that $w(K) = K_w^{*f}$ then $S'_{f,w}$ holds in K if and only if S holds in $K/w(K)$. In particular, the sentence

$$I(G/w(G))'_{f,w}$$

is true in G. So it is true in H, from which it follows that $H/w(H) \cong G/w(G)$.

Now let N be any open normal subgroup of H. By Exercise 4.2.5 there exists a d-locally finite word w with $w(H) \le N$, and we obtain an epimorphism

$$G \to G/w(G) \cong H/w(H) \to H/N;$$

it is continuous because its kernel contains the open subgroup $w(G)$. Since the profinite group H is the inverse limit of all such quotients H/N, it follows that H is an epimorphic image of G. Let $\theta : G \to H$ be an epimorphism, and suppose that $1 \ne g \in G$. Then $g \notin M$ for some open normal subgroup M of G, and as before we can find a d-locally finite word w such that $w(G) \le M$. Now θ induces an epimorphism $\theta^* : G/w(G) \to H/w(H)$; as these groups are isomorphic and finite, θ^* is an isomorphism. It follows that

$$g\theta \cdot w(H) = (g \cdot w(G))\theta^* \ne 1$$

since $g \notin w(G)$. Thus $g\theta \ne 1$, so θ is injective and hence an isomorphism $G \to H$.

Exercise 4.2.6. Let G be a profinite group and N a normal subgroup of finite index. Show that G/N is isomorphic to a section of G/T for some $T \lhd_o G$. [*Hint*: note that $G = NH$ for some finitely generated closed subgroup H of G.]

4.3   Pronilpotent groups

A remarkably explicit characterization of the words that are uniformly elliptic in finite p-groups, or in finite nilpotent groups, was discovered by Andrei Jaikin. We will combine his results and determine which words are uniformly elliptic in the class $\mathfrak{N}(\pi)$ of all finite nilpotent π-groups, where π is an arbitrary set of primes.

Let $\mathcal{C}$ be a formation consisting of finite nilpotent groups; we will denote by E_k or E the free pro-$\mathcal{C}$ group on k generators $x_1, \ldots, x_k$. Any element $u \in E = E_k$ may be construed as a 'pro-$\mathcal{C}$ word', and 'evaluated' in an arbitrary pro-$\mathcal{C}$ group G: given $\mathbf{g} = (g_1, \ldots, g_k) \in G^{(k)}$ we set

$$u(\mathbf{g}) = u\pi_{\mathbf{g}}$$

where $\pi_{\mathbf{g}} : E \to G$ is the unique continuous homomorphism $E \to G$ sending x_i to g_i for $i = 1, \ldots, k$. Recall that E is the pro-$\mathcal{C}$ completion of the free group F_k on $\{x_1, \ldots, x_k\}$; thus the natural image $X = \langle x_1, \ldots, x_k \rangle$ of F_k in E is dense in E, so $E = XM$ for every open normal subgroup M of E. Given such an M, there exists an ordinary word $v = v(x_1, \ldots, x_k)$ such that $u \equiv v \mod M$, and then for $\mathbf{g} = (g_1, \ldots, g_k) \in G^{(k)}$ we have

$$u(\mathbf{g}) = u\pi_{\mathbf{g}} \equiv v\pi_{\mathbf{g}} = v(g_1, \ldots, g_k) \mod M\pi_{\mathbf{g}}.$$

Since $\pi_{\mathbf{g}}$ is a continuous homomorphism, given any $N \lhd_o G$ we can find $M \lhd_o E$ with $M\pi_{\mathbf{g}} \leq N$. Thus for any such N we can determine $u(\mathbf{g})$ modulo N by evaluating an ordinary word in the usual way. (*Remark*: I am slightly abusing notation here, by blurring the distinction between $x_i \in E$ and $x_i \in F_k$. When F_k is residually a $\mathcal{C}$ group, the natural map $\iota : F_k \to X$ is an isomorphism and X can be identified with F_k; in general, the kernel of ι is the $\mathcal{C}$-residual of F_k. In any case $\pi_{\mathbf{g}} : E \to G$ composes with ι to give the evaluation mapping $\pi_{\mathbf{g}} : F_k \to G$.)

Of course, the preceding observations do not require $\mathcal{C}$ to consist of nilpotent groups; most of the following results do. We begin by re-interpreting in profinite terms some basic combinatorial results from Section 1.2.

Lemma 4.3.1 *Let* $G = \overline{\langle y_1, \ldots, y_s \rangle}$ *be a pronilpotent group, and put* $G_l = \gamma_l(G)$ *for each* l. *Let* $n, c \in \mathbb{N}$.

(i) *For each* $q, l \in \mathbb{N}$ *the subgroup* $G^q G_l$ *is open in* G.

(ii)

$$G_{c+n} = \prod [G_n, y_{i_1}, \ldots, y_{i_c}], \tag{4.3}$$

the product ranging over $\mathbf{i} = (i_1, \ldots, i_c) \in [1, s]^{(c)}$, *in any chosen order.*

(iii) *Let* $g \in G$ *and* $h \in G_n$. *Then* $[h, y_{i_1}, \ldots, y_{i_c}]$ *is central in* G *modulo* G_{n+c+1} *and*

$$[gh, y_{i_1}, \ldots, y_{i_c}] \equiv [g, y_{i_1}, \ldots, y_{i_c}][h, y_{i_1}, \ldots, y_{i_c}] \mod G_{n+c+1}.$$

(iv) *Let* $u \in \gamma_r(\eta(E))$, *where* η *is any word such that* $\eta(G)$ *is open in* G. *Then for* $\mathbf{g} = (g_1, \ldots, g_k) \in G^{(k)}$ *and* $h \in \gamma_n(\eta(G))$ *we have*

$$u(g_1 h, g_2, \ldots, g_k) \equiv u(\mathbf{g}) \mod \gamma_{n+r-1}(\eta(G)).$$

Proof. (i) Let

$$\mathcal{N} = \{N \lhd_o G \mid G^q G_l \leq N\}.$$

If $N \in \mathcal{N}$ then G/N is an image of the relatively free group $F_s/\gamma_l(F_s)F_s^q$, which is finite. It follows that $\mathcal{N}$ has a unique minimal member N_0, say. Now the verbal subgroup $G^q G_l$ is closed in G, by Corollary 4.1.6, so

$$G^q G_l = \bigcap \mathcal{N} = N_0 \lhd_o G.$$

(ii) Let Q denote the expression on the right-hand side of (4.3). Let $N \lhd_o G$. Then G/N is generated by the images of $y_1, \ldots, y_s$; applying Proposition 1.2.7 to the nilpotent group G/N we get

$$G_{c+n} N = QN.$$

Now Theorem 4.1.5(i) shows that both G_{c+n} and Q are closed in G (the latter being a finite product of closed sets). Therefore

$$G_{c+n} = \bigcap_{N \lhd_o G} G_{c+n} N = \bigcap_{N \lhd_o G} QN = Q.$$

Part (iii) is deduced in a similar way from Proposition 1.2.1. For (iv), write $\mathbf{h} = (g_1 h, g_2, \ldots, g_k)$. Let $N \lhd_o G$, and choose $M \lhd_o E$ so that each of the continuous homomorphisms $\pi_{\mathbf{g}}$, $\pi_{\mathbf{h}}$ maps M into N. As $u \in \gamma_r(\eta(E))$, there exists $v \in \gamma_r(\eta(F_k))$ such that $u \equiv v(x_1, \ldots, x_k) \mod M$. Corollary 1.2.4 shows that

$$v(\mathbf{h}) \equiv v(\mathbf{g}) \mod \gamma_{n+r-1}(\eta(G)).$$

As $u(\mathbf{h}) \equiv v(\mathbf{h})$ and $u(\mathbf{g}) \equiv v(\mathbf{g}) \mod N$ it follows that

$$u(\mathbf{h}) \equiv u(\mathbf{g}) \mod N \gamma_{n+r-1}(\eta(G)).$$

Thus

$$u(\mathbf{h})^{-1} u(\mathbf{g}) \in \bigcap_{N \lhd_o G} \gamma_{n+r-1}(\eta(G))N = \gamma_{n+r-1}(\eta(G))$$

as $\gamma_{n+r-1}(\eta(G))$ is closed (note that $\eta(G)$ is a finitely generated pronilpotent group because it is open in G). ■

The special case $c = n = 1$ of (ii) will frequently be used without special mention:

Corollary 4.3.2 *If $G = \overline{\langle y_1, \ldots, y_s \rangle}$ is pronilpotent then*

$$G' = \overline{G'} = [G, y_1][G, y_2] \ldots [G, y_s].$$

We are now ready to prove the following generalization of Corollary 4.1.6:

Theorem 4.3.3 [J1] *Let w be a word, let $d \in \mathbb{N}$ and suppose that $E/\overline{w(E)}$ is virtually nilpotent where $E = E_{d+1}$ is the free pro-$\mathcal{C}$ group on $d + 1$ generators. Then $w(G)$ is closed in G for every d-generator pro-$\mathcal{C}$ group G.*

Proof. Label the free generators of E as $z, x_1, \ldots, x_d$. We shall show that if $G = \overline{\langle x_1, \ldots, x_d \rangle}$ then $w(G)$ is closed in G. This will suffice to establish the theorem, since every d-generator pro-$\mathcal{C}$ group is an epimorphic image of G (and the image of a closed – hence compact – set is again compact, hence closed).

There exist $H \lhd_o E$ and $c \in \mathbb{N}$ such that $\gamma_c(H) \leq w(E)$. Then E/H is a finite nilpotent group (because it is in $\mathcal{C}$), so $H \geq E^q \gamma_l(E) = \eta(E)$ for some $q, l \in \mathbb{N}$, where η is the word $[x_1, \ldots, x_l]x_{l+1}^q$. Also $\eta(E) \lhd_o E$ by Lemma 4.3.1(i), so changing notation we may as well assume that $H = \eta(E)$. Write $H_n = \gamma_n(H)$, $G_n = \gamma_n(G)$ for each n.

Now Theorem 4.1.5(i), (ii) shows that for each n, the word w has finite width $\mu(n)$, say, in the finitely generated virtually nilpotent profinite group E/H_n, and that $w(E/H_n) = w(E)H_n/H_n$ is closed. Therefore $w(E)H_n$ is closed in E, and taking $m = \mu(c + 2)$ we thus have, in particular,

$$H_c \leq \overline{w(E)} \leq w(E)H_{c+2} = E_w^{*m} H_{c+2}. \tag{4.4}$$

Now $\eta(G)$ is open in G so it is finitely generated; say

$$\eta(G) = \overline{\langle y_1, \ldots, y_s \rangle}.$$

For $\mathbf{i} \in [1, s]^{(c)}$ write

$$[z, \mathbf{y}(\mathbf{i})] = [z, y_{i_1}, \ldots, y_{i_c}].$$

Since $[z, y_{i_1}] \in H$ this element lies in H_c, so there exist $v_{\mathbf{i}} \in E_w^{*m}$ and $u_{\mathbf{i}} \in H_{c+2} = \gamma_{c+2}(\eta(E))$ such that

$$[z, \mathbf{y}(\mathbf{i})] = v_{\mathbf{i}} u_{\mathbf{i}}. \tag{4.5}$$

We will prove by induction on n that

$$\gamma_{c+1}(\eta(G)) \subseteq \prod_{\mathbf{i} \in [1,s]^{(c)}} v_{\mathbf{i}}(\eta(G), x_1, \ldots, x_d) \cdot \gamma_{c+n}(\eta(G)) \tag{4.6}$$

for every $n \geq 1$ (with some arbitrary, but fixed, ordering of $[1, s]^{(c)}$). Here, $v_{\mathbf{i}}$ is playing its role as a pro-$\mathcal{C}$ word, so $v_{\mathbf{i}}(\eta(G), x_1, \ldots, x_d)$ denotes the set $\{v_{\mathbf{i}}\pi_g \mid g \in \eta(G)\}$ where $\pi_g : E \to G$ maps z to g and fixes $x_1, \ldots, x_d$. Thus $v_{\mathbf{i}}\pi_g \in G_w^{*m}$ for each $g \in G$, so (4.6) implies that

$$\gamma_{c+1}(\eta(G)) \subseteq G_w^{*ms^c} \cdot \gamma_{c+n}(\eta(G)).$$

Now if $N \lhd_o G$ then $N \geq \gamma_{c+n}(\eta(G))$ for sufficiently large n. So if (4.6) holds for every n then

$$\gamma_{c+1}(\eta(G)) \subseteq \bigcap_{N \lhd_o G} G_w^{*ms^c} \cdot N = G_w^{*ms^c} \subseteq w(G)$$

since $G_w^{*ms^c}$ is a closed set. Then Corollary 4.1.6 shows that $w(G)$ is closed in G, as required.

The claim (4.6) is trivial for $n = 1$. Now suppose that (4.6) is true for some $n \geq 1$. Let $h \in \gamma_{c+1}(\eta(G))$. Then, in view of Lemma 4.3.1(ii), there exist $g_i \in \eta(G)$ and $f_i \in \gamma_n(\eta(G))$ such that

$$h = \prod_i v_i(g_i, x_1, \ldots, x_d) \cdot \prod_i [f_i, \mathbf{y}(i)]$$
$$= \prod_i [g_i, \mathbf{y}(i)] u_i(g_i, x_1, \ldots, x_d)^{-1} \cdot \prod_i [f_i, \mathbf{y}(i)],$$

where i runs through $[1, s]^{(c)}$ in each product. Now each of the terms $u_i(g_i, x_1, \ldots, x_d)$, $[g_i, \mathbf{y}(i)]$ lies in $\eta(G)$, while $[f_i, \mathbf{y}(i)]$ is central in $\eta(G)$ modulo $\gamma_{c+n+1}(\eta(G))$. Also, since $u_i \in \gamma_{c+2}(\eta(E))$, Lemma 4.3.1(iv) shows that

$$u_i(g_i, x_1, \ldots, x_d)^{-1} \equiv u_i(g_i f_i, x_1, \ldots, x_d)^{-1} \bmod \gamma_{c+n+1}(\eta(G))$$
$$= [g_i f_i, \mathbf{y}(i)]^{-1} v_i(g_i f_i, x_1, \ldots, x_d)$$

for each i. It follows that

$$h \equiv \prod_i [g_i, \mathbf{y}(i)][f_i, \mathbf{y}(i)] u_i(g_i, x_1, \ldots, x_d)^{-1}$$
$$\equiv \prod_i [g_i, \mathbf{y}(i)][f_i, \mathbf{y}(i)][g_i f_i, \mathbf{y}(i)]^{-1} v_i(g_i f_i, x_1, \ldots, x_d)$$
$$\equiv \prod_i v_i(g_i f_i, x_1, \ldots, x_d) \bmod \gamma_{c+n+1}(\eta(G)),$$

using Lemma 4.3.1(iii) in the final step (with $\eta(G)$ in place of G).

Thus (4.6) holds with $n + 1$ in place of n. It therefore holds for all n by induction, and this completes the proof. ∎

Let us try to identify which words w satisfy the hypothesis of Theorem 4.3.3. Let p be a prime and let $w \in F = F_k$. In the following section we will establish the equivalence of the following three conditions:

1. $w(C_p \wr C_{p^n}) = 1$ for infinitely many n;

2. $w(C_p \wr C_\infty) = 1$;

3. $w \in F''(F')^p$.

Definition The word w is a *J(p) word* if it does *not* satisfy the conditions **1–3**. For any set of primes π, we say that w is a *J(π) word* if w is a J(p) word for each $p \in \pi$.

Exercise 4.3.1. Show that $[x_1^q, \ldots, x_c^q]$ is a J(p) word for each $c, q \in \mathbb{N}$. [*Hint*: identify $C_p \wr C_\infty$ with $\mathbb{F}_p \langle x \rangle \rtimes \langle x \rangle$ and evaluate $[(a \cdot x)^q, x^q, \ldots, x^q]$ where $a = 1_{\mathbb{F}_p \langle x \rangle}$.]

Theorem 4.3.4 *The following are equivalent:*

(a) w *is a* $J(p)$ *word;*

(b) $E/\overline{w(E)}$ *is virtually nilpotent where* E *is the free pro-p group on* 2 *generators;*

(c) $G/\overline{w(G)}$ *is virtually nilpotent for every finitely generated pro-p group* G.

Proof. Obviously (c) implies (b). Suppose that (b) holds; then for some positive integers q and c we have $\gamma_c(E^q) \leq \overline{w(E)}$. Now E maps onto $C_p \wr C_{p^n}$ for every n; if $w(C_p \wr C_{p^n}) = 1$ then $E/\overline{w(E)}$ maps onto $C_p \wr C_{p^n}$. This implies that $\gamma_c((C_p \wr C_{p^n})^q) = 1$, which is false for large enough n by the preceding exercise. Thus (b) implies (a).

To show that (a) implies (c) we may as well assume that G is a free pro-p group. Put $P = G/\overline{w(G)}$. If $C_p \wr C_{p^n}$ is a closed section of P then $w(C_p \wr C_{p^n}) = 1$; so (a) implies that $C_p \wr C_{p^n}$ is not a closed section of P if n is large. It now follows by a theorem of Shalev [S10] that P is p-adic analytic, i.e. that P has finite rank as a pro-p group (see [DDMS], Exercise 3.4). The next proposition shows that in this case, P is virtually nilpotent, as required. ∎

Proposition 4.3.5 [BM] *Let* w *be a group word and* G *a finitely generated free pro-p group. If* $P = G/\overline{w(G)}$ *has finite rank then* P *is virtually nilpotent.*

Proof. We may assume that G is non-abelian; then certainly G has infinite rank, so w must be a non-trivial word. Now P is a linear group over $\mathbb{Q}_p$ (see [DDMS]); a theorem of Platonov (see [W1], **10.15**) says that a linear group satisfying a non-trivial identity is virtually soluble (of course this follows from the well-known 'Tits alternative', but that is a much harder result). Thus P is virtually soluble.

It follows that P has closed normal subgroups $N \leq H$ such that P/H is finite, H/N is abelian and N is torsion-free and nilpotent (by the Lie-Kolchin-Mal'cev theorem; see [W1], Chapter 3). Let Z_i denote the ith term of the upper central series of N; then each factor Z_i/Z_{i-1} is a torsion-free abelian pro-p group of rank at most r, the rank of P. Writing Z_i/Z_{i-1} additively we consider

$$V_i := (Z_i/Z_{i-1}) \otimes \mathbb{Q}_p$$

as a $\mathbb{Q}_p H$-module, where H acts by conjugation. Note that V_i is a $\mathbb{Q}_p$-vector space of dimension at most r. I claim that for each i and each $h \in H$,

- $h^{r!}$ acts unipotently on V_i.

As H/N is abelian, this will imply that the group $NH^{r!}$ is nilpotent, and the result follows, because $\overline{NH^{r!}}$ is then also nilpotent, and has finite index in P (in fact $NH^{r!}$ is already closed, cf. [DDMS], Chapter 1).

To establish the claim, let $a \in Z_i/Z_{i-1}$, put $U = a \cdot \mathbb{Q}_p \langle h \rangle \leq V_i$ and let $\eta \in \mathrm{GL}(U)$ correspond to the action of h. As V_i is the sum of finitely many modules like U, it will suffice to show that $\eta^{r!}$ is unipotent. Now consider the two elements

$$v = (a, \ldots, a), \ \gamma = (\gamma_1, \ldots, \gamma_n) \in (P/Z_{i-1})^{(n)},$$

the direct product of $n = r + 1$ copies of P/Z_{i-1}, where $\gamma_j = h^j Z_{i-1}$ for each j. Put $\Delta = \overline{\langle v, \gamma \rangle}$. Since P is relatively free on at least two generators, there is an epimorphism from P onto Δ, and so Δ has rank at most r. Therefore

$$W := \left((Z_i/Z_{i-1})^{(n)} \cap \Delta \right) \otimes \mathbb{Q}_p$$

is a $\mathbb{Q}_p$-vector space of dimension at most r. Evidently

$$v^\gamma = (a\eta, a\eta^2, \ldots, a\eta^n).$$

Let $f(X)$ denote the characteristic polynomial of the action of γ on W. Then $f(\gamma)$ annihilates W, and as $v \in W$ it follows that for $j = 1, \ldots, n$ we have $a \cdot f(\eta^j) = 0$, whence $f(\eta^j) = 0$. Now let λ be an eigenvalue of η. As f has at most $r < n$ roots, there exist s and t with $1 \le s < t \le n$ such that $\lambda^s = \lambda^t$. Thus $\lambda^e = 1$ where $e = t - s \le n - 1 = r$, and so $\lambda^{r!} = 1$.

Thus $\eta^{r!}$ is unipotent as claimed. ∎

(This argument is taken from [S5], where several other results of a similar nature may be found.)

Theorem 4.3.6 *Let π be a non-empty set of primes, and let w be a $J(\pi)$ word. Then w is uniformly elliptic in $\mathfrak{N}(\pi)$.*

Proof. We have to show that $w(G)$ is closed in G for every finitely generated pro-$\mathcal{C}$ group G, where $\mathcal{C} = \mathfrak{N}(\pi)$. This will follow by Theorem 4.3.3 if we prove that $E/w(E)$ is virtually nilpotent for each finitely generated free pro-$\mathcal{C}$ group E.

Fix $m \in \mathbb{N}$, set $F = F_m$ and let $K/w(F)$ be the $\mathcal{C}$-residual of $F/w(F)$, i.e.

$$K = \bigcap \{ N \lhd F \mid w(F) \le N, \ F/N \in \mathfrak{N}(\pi) \}.$$

Put $\Gamma = F/K$.

For each $p \in \pi$, the pro-p completion $\widehat{\Gamma}_p$ of Γ is an m-generator pro-p group, and satisfies $w(\widehat{\Gamma}_p) = 1$; it follows by Theorem 4.3.4 that $\widehat{\Gamma}_p$ is virtually nilpotent. Now Proposition 4.3.7, proved below, shows that $\widehat{\Gamma}_\mathcal{C}$ is virtually nilpotent.

The free pro-$\mathcal{C}$ group $E = E_m$ on m generators is the pro-$\mathcal{C}$ completion of F. Therefore

$$E/\overline{w(E)} = \varprojlim \{ F/N \mid w(F) \le N \lhd F, \ F/N \in \mathfrak{N}(\pi) \} = \widehat{\Gamma}_\mathcal{C},$$

and the result follows. ∎

Proposition 4.3.7 *Let Γ be a finitely generated group and π a non-empty set of primes. Suppose that $\widehat{\Gamma}_p$ is virtually nilpotent for each $p \in \pi$. Then $\widehat{\Gamma}_{\mathfrak{N}(\pi)}$ is virtually nilpotent, and if Γ is residually $\mathfrak{N}(\pi)$ then Γ is virtually nilpotent.*

Proof. Choose $p_1 \in \pi$, and let Γ_{p_1} denote the $\mathfrak{F}_{p_1}$-residual of Γ; this is the kernel of the natural map $\Gamma \to \widehat{\Gamma}_{p_1}$. Then Γ/Γ_{p_1} is virtually nilpotent, so it is a polycyclic group.

Now suppose that Γ/T is a torsion-free nilpotent quotient of Γ. Then Γ/T is residually a finite p_1-group ([S3], Chapter 1, Theorem 4), so $T \geq \Gamma_{p_1}$. Hence the Hirsch length $h(\Gamma/T)$ of Γ/T is bounded above by $h(\Gamma/\Gamma_{p_1})$. We may therefore choose T so as to maximize $h(\Gamma/T)$. In this case, I claim that $T/[T,\Gamma]$ *is finite*. To see this, let $S/[T,\Gamma]$ denote the torsion subgroup of $\Gamma/[T,\Gamma]$. Then $\Gamma/T \cap S$ is torsion-free and nilpotent, so

$$h(\Gamma/T) + h(T/T \cap S) = h(\Gamma/T \cap S) \leq h(\Gamma/T)$$

which shows that $h(T/T \cap S) = 0$; as $T/T \cap S$ is torsion-free we must have $T \cap S = T$, whence $T/[T,\Gamma] \leq S/[T,\Gamma]$, a finite group.

Let σ denote the set of prime divisors of $|T/[T,\Gamma]|$. Let q be any prime with $q \notin \sigma$, and suppose that Γ/Q is a quotient of Γ that is a finite q-group. Then

$$\frac{TQ}{[T,\Gamma]Q} \cong \frac{T}{[T,\Gamma](T \cap Q)} = 1,$$

since the subgroups $[T,\Gamma]$ and $T \cap Q$ have coprime indices in T. As Γ/Q is nilpotent this implies that $TQ = Q$.

Thus writing c for the nilpotency class of Γ/T we have

$$\gamma_{c+1}(\Gamma) \leq T \leq Q$$

whenever Γ/Q is a finite q-group for a prime $q \notin \sigma$.

Now the pro-$\mathfrak{N}(\pi)$ completion of Γ is

$$\widehat{\Gamma}_{\mathfrak{N}(\pi)} = \prod_{p \in \pi} \widehat{\Gamma}_p = \prod_{p \in \pi \cap \sigma} \widehat{\Gamma}_p \times \prod_{q \in \pi \setminus \sigma} \widehat{\Gamma}_q.$$

It follows from the preceding paragraph that each of the factors $\widehat{\Gamma}_q$ with $q \in \pi \setminus \sigma$ is nilpotent of class at most c; therefore so is $\prod_{q \in \pi \setminus \sigma} \widehat{\Gamma}_q$. Each factor $\widehat{\Gamma}_p$ with $p \in \pi$ is virtually nilpotent by hypothesis; as $\pi \cap \sigma$ is a finite set, it follows that $\widehat{\Gamma}_{\mathfrak{N}(\pi)}$ is virtually nilpotent. If Γ is residually $\mathfrak{N}(\pi)$ then Γ embeds in $\widehat{\Gamma}_{\mathfrak{N}(\pi)}$, whence the second claim of the proposition. ∎

Remark. Further applications of the above argument are given in Theorem 8 of [LS3], Window 8; this shows that if the upper p-ranks of a finitely generated residually nilpotent group are all finite, then they are uniformly bounded over all primes p.

The *converse* of Theorem 4.3.6 is also true; this will be established in §4.5.

One can always recognize a J(π) word by a simple rewriting process:

Exercise 4.3.2. Show that F_k' is free on the generating set

$$\left\{ y(i, \mathbf{e}) = [x_1^{e_1} \ldots x_i^{e_i}, x_i]^{x_{i+1}^{e_{i+1}} \ldots x_k^{e_k}} \mid 2 \leq i \leq k, \ \mathbf{e} \in E_i \right\}$$

where $E_i = \left\{ \mathbf{e} \in \mathbb{Z}^{(k)} \mid e_j \neq 0 \text{ for some } j < i \right\}$. Deduce that if $w \in F_k'$ then

$$w \equiv \prod_{i=2}^{k} \prod_{\mathbf{e} \in E_i} y(i, \mathbf{e})^{n(i, \mathbf{e})} \mod F_k''$$

for uniquely determined integers $n(i, \mathbf{e})$; in this case define $n(w) = \gcd_{i, \mathbf{e}} \{n(i, \mathbf{e})\}$. Deduce that w is a J(π) word if and only if either $w \notin F_k'$ or $w \in F_k'$ and $p \nmid n(w)$ for every $p \in \pi$.

4.4 Variety stuff

For a group variety $\mathcal{V}$ we set

$$F_d(\mathcal{V}) = F_d / \mathcal{V}(F_d);$$

this is the free $\mathcal{V}$-group on d generators. Here $d \in \mathbb{N}$ or $d = \infty$ (meaning $d = \aleph_0$).

Lemma 4.4.1 $F_\infty(\mathcal{V})$ *is residually* $\{F_d(\mathcal{V}) \mid d \in \mathbb{N}\}$.

Proof. Let $g \in F_\infty \smallsetminus \mathcal{V}(F_\infty)$. Then $g \in F_d = \langle x_1, \ldots, x_d \rangle$ for some finite d. Let $\pi : F_\infty \to F_d$ be the epimorphism sending x_i to x_i for $1 \leq i \leq d$ and x_j to 1 for $j > d$. Then π induces an epimorphism $\overline{\pi} : F_\infty(\mathcal{V}) \to F_d(\mathcal{V})$. If $g \in \ker \overline{\pi}$ then $g\pi \in \mathcal{V}(F_d) \leq \mathcal{V}(F_\infty)$; but $g\pi = g \notin \mathcal{V}(F_\infty)$. The lemma follows. ∎

Proposition 4.4.2 *Let* $\mathfrak{X}$ *be a subgroup-closed family of groups and let* $\mathcal{V}$ *be the variety generated by* $\mathfrak{X}$. *Let* $d \in \mathbb{N}$. *Then* $F_d(\mathcal{V})$ *is residually-*$\mathfrak{X}$.

Proof. Let A be the Cartesian product of all groups in $\mathfrak{X}$ and let

$$A^* = A^{A^{(d)}}.$$

For $i = 1, \ldots, d$ define $y_i \in A^*$ by

$$y_i(a_1, \ldots, a_d) = a_i \quad (a_1, \ldots, a_d \in A),$$

and set $Y = \langle y_1, \ldots, y_d \rangle \leq A^*$. I claim that $Y \cong F_d(\mathcal{V})$. This implies the result since $\mathfrak{X}$ is subgroup-closed and A^* is clearly residually-$\mathfrak{X}$.

To establish the claim, it will suffice to show that the kernel K of the homomorphism $\pi : F_d \to Y$ sending x_i to y_i for each i is exactly $\mathcal{V}(F_d)$. Certainly $\mathcal{V}(F_d) \leq K$ since $\mathcal{V}(F_d)\pi = \mathcal{V}(Y) \leq \mathcal{V}(A^*) = 1$. Now suppose $w \in F_d \smallsetminus \mathcal{V}(F_d)$. Then $w(H) \neq 1$ for some $H \in \mathfrak{X}$, so $w(A) \neq 1$. Thus there exists $\mathbf{a} \in A^{(d)}$ such that $w(\mathbf{a}) \neq 1$. Then

$$(w\pi)(\mathbf{a}) = w(y_1(\mathbf{a}), \ldots, y_d(\mathbf{a})) = w(a_1, \ldots, a_d) \neq 1$$

so $w \notin K$. The claim follows. ∎

Proposition 4.4.3 *Let $\mathfrak{X}$ be a subgroup-closed class of groups and let $\mathcal{V}$ be a variety that contains $\mathfrak{X}$. Then the following are equivalent:*

(a) *$\mathcal{V}$ is generated by groups in $\mathfrak{X}$;*

(b) *$\mathcal{V}$ is generated by groups that are residually-$\mathfrak{X}$;*

(c) *every finitely generated free $\mathcal{V}$-group is residually-$\mathfrak{X}$;*

(d) *$F_\infty(\mathcal{V})$ is residually-$\mathfrak{X}$.*

Proof. Trivially (a) implies (b). Suppose now that (b) holds. Let $\mathfrak{X}'$ be a subgroup-closed family of residually-$\mathfrak{X}$ groups that generates $\mathcal{V}$. The preceding proposition shows that $F_d(\mathcal{V})$ is residually-$\mathfrak{X}'$, hence residually-$\mathfrak{X}$, for each finite d. Thus (c) holds, and this implies (d) by Lemma 4.4.1. Suppose finally that (d) holds. If a word $w = w(x_1, \ldots, x_k)$ satisfies $w(H) = 1$ for every $H \in \mathfrak{X}$ then $w(F_\infty(\mathcal{V})) = 1$, so $w \in \mathcal{V}(F_\infty)$. If $G \in \mathcal{V}$ and $g_1, \ldots, g_k \in G$ then $\langle g_1, \ldots, g_k \rangle$ is a homomorphic image of $F_\infty(\mathcal{V})$, so $w(g_1, \ldots, g_k) = 1$. Thus $w(G) = 1$. Thus the variety generated by $\mathfrak{X}$ contains $\mathcal{V}$, and (a) follows since $\mathfrak{X}$ is contained in $\mathcal{V}$. ∎

Definition For two varieties $\mathcal{A}$, $\mathcal{B}$, the *product variety $\mathcal{AB}$* consists of all groups G having a normal subgroup N with $N \in \mathcal{A}$ and $G/N \in \mathcal{B}$.

Exercise 4.4.1. Show that $\mathcal{AB}$ is a variety. [*Hint:* put $B = \mathcal{B}(F_\infty)$ and $A = \mathcal{A}(B)$. Show that $G \in \mathcal{AB}$ if and only if $w(G) = 1$ for every $w \in A$.]

Lemma 4.4.4 *Let $\mathcal{A}$ and $\mathcal{B}$ be varieties. Then the variety $\mathcal{AB}$ is generated by the family of groups*

$$\{F_d(\mathcal{A}) \wr F_d(\mathcal{B}) \mid d \in S\}$$

for any infinite set S of positive integers.

Proof. Let A, B be groups and w a word such that $w(A \wr B) = 1$. I claim that then $w(A \bar{\wr} B) = 1$, where $A \bar{\wr} B$ denotes the complete wreath product of A and B. To see this, write

$$A \bar{\wr} B = \overline{U} \rtimes B$$

where $\overline{U} = A^B$ is the base group. Suppose that $w(a_1 y_1, \ldots, a_k y_k) \neq 1$ with $a_i \in \overline{U}$ and $y_i \in B$ for each i. Since $w(B) = 1$, we have

$$1 \neq w(a_1 y_1, \ldots, a_k y_k) = w'_{\mathbf{y}}(\mathbf{a}) \in \overline{U}.$$

Then $w'_{\mathbf{y}}(\mathbf{a})\pi_z \neq 1$ for some $z \in B$, where $\pi_z : \overline{U} \to A$ denotes the z-co-ordinate projection. Now

$$w'_{\mathbf{y}}(\mathbf{a}) = \prod_{j=1}^{s} a_{i_j}^{\varepsilon_j v_j}$$

where $v_1, \ldots, v_s \in B$ and each ε_j is ± 1. As B acts on $\overline{U}$ by permuting the components, it follows that

$$w'_{\mathbf{y}}(\mathbf{a})\pi_z = w'_{\mathbf{y}}(\widetilde{\mathbf{a}})\pi_z$$

provided $\widetilde{a}_i\pi_y = a_i\pi_y$ for each $y \in \{zv_1^{-1}, \ldots, zv_s^{-1}\}$ and each i. We can choose each $\widetilde{a}_i$ so that $\widetilde{a}_i\pi_y = 1$ for every $y \notin \{zv_1^{-1}, \ldots, zv_s^{-1}\}$. Then each $\widetilde{a}_iy_i$ lies in the restricted wreath product $A \wr B$; so $w(\widetilde{\mathbf{a}} \cdot \mathbf{y}) = 1$. This implies

$$1 = w(\widetilde{\mathbf{a}} \cdot \mathbf{y})\pi_z = w'_{\mathbf{y}}(\mathbf{a})\pi_z,$$

a contradiction which establishes the claim.

Now let $G \in \mathcal{AB}$ and let w be a word such that $w(G) \neq 1$. There exists $A \lhd G$ such that $A \in \mathcal{A}$ and $G/A = B \in \mathcal{B}$. According to the Kaloujnine-Krasner embedding theorem ([N], Theorem 22.21), G is isomorphic to a subgroup of the complete wreath product $W = A\bar{\wr}B$; so $w(W) \neq 1$, and by the first paragraph it follows that $w(A \wr B) \neq 1$. Then $w(A_0 \wr B_0) \neq 1$ for some finitely generated subgroups A_0 and B_0 of A, B respectively. If d is large enough, A_0 is an image of $F_d(\mathcal{A})$ and B_0 is an image of $F_d(\mathcal{B})$, so $F_d(\mathcal{A}) \wr F_d(\mathcal{B})$ maps onto $A_0 \wr B_0$. We may conclude that $w(F_d(\mathcal{A}) \wr F_d(\mathcal{B})) \neq 1$. The result follows. $\blacksquare$

We will write $\mathfrak{A}$ to denote the variety of all abelian groups, and $\mathfrak{A}_p$ to denote the variety of all elementary abelian groups of exponent p. Thus $\mathfrak{A}_p\mathfrak{A}$ is the class of all groups G whose derived group G' is an elementary abelian p-group; as a variety it is defined by the laws

$$[[x_1, x_2], [x_3, x_4]] = 1 = [x_1, x_2]^p.$$

The relatively free group on k generators in this variety is $F/F''(F')^p$ where $F = F_k$. So $w \in F''(F')^p$ if and only $w(G) = 1$ for every $G \in \mathfrak{A}_p\mathfrak{A}$.

Proposition 4.4.5 *The variety $\mathfrak{A}_p\mathfrak{A}$ is generated by the family of groups $\mathfrak{X}(S) = \{C_p \wr C_{p^n} \mid n \in S\}$ for any infinite set S of positive integers.*

Proof. Let $\langle x \rangle$ be an infinite cyclic group and let $R = \mathbb{F}_p \langle x \rangle$ be its group algebra over $\mathbb{F}_p$. Then $C_p \wr C_\infty = R \rtimes \langle x \rangle$ where R is considered as an $\langle x \rangle$-module. Write $I = (x-1)R$ for the augmentation ideal of R. Then $I^{p^n} = (x^{p^n} - 1)R$ for $n \in \mathbb{N}$, so R/I^{p^n} may be identified with the group algebra of $\langle x \rangle / \langle x^{p^n} \rangle$ and we have

$$\frac{R \rtimes \langle x \rangle}{I^{p^n} \cdot \langle x^{p^n} \rangle} \cong \frac{R}{I^{p^n}} \rtimes \frac{\langle x \rangle}{\langle x^{p^n} \rangle} = C_p \wr C_{p^n}.$$

As S is infinite we have $\bigcap_{n \in S} I^{p^n} = 0$ (every non-zero ideal of R has finite index!); consequently

$$\bigcap_{n \in S} I^{p^n} \cdot \langle x^{p^n} \rangle = 1.$$

Thus $C_p \wr C_\infty$ is residually $\mathfrak{X}(S)$.

Now let $n \in \mathbb{N}$ and let $A_n = \langle a_1, \ldots, a_n \rangle$ denote the free abelian group of rank n. I claim that $C_p \wr A_n$ is residually $C_p \wr C_\infty$. To see this, let f be a non-trivial element of the base group in $C_p \wr A_n$. Identifying the base group with $\mathbb{F}_p A_n$, we can write $f = f(a_1, \ldots, a_n)$ where

$$f(X_1, \ldots, X_n) = \sum \lambda(\mathbf{e}) X_1^{e_1} \ldots X_n^{e_n}$$

with $0 \neq \lambda(\mathbf{e}) \in \mathbb{F}_p$ for each $\mathbf{e}$. *Subclaim:* there exist $c_1, \ldots, c_n \in \mathbb{Z}$, with $\gcd(c_1, \ldots, c_n) = 1$, such that $f(X^{c_1}, \ldots, X^{c_n}) \neq 0$. Indeed, it suffices to find $\mathbf{c} \in \mathbb{Z}^n$ such that the finitely many numbers $\mathbf{c} \cdot \mathbf{e} = c_1 e_1 + \cdots + c_n e_n$ are pairwise distinct. The set of $\mathbf{c} \in \mathbb{Q}^n$ which fail this test is the union of finitely many subspaces of codimension one in $\mathbb{Q}^n$, so its complement is infinite. Take any member of this complement and clear denominators to obtain $\mathbf{c} \in \mathbb{Z}^n$ with the required property.

Let $\langle x \rangle$ be an infinite cyclic group. Given the *subclaim*, we see that the mapping

$$\phi_1 : A_n \to \langle x \rangle$$
$$a_i \mapsto x^{c_i} \quad (i = 1, \ldots, n)$$

induces an epimorphism $\phi_2 : \mathbb{F}_p A_n \to \mathbb{F}_p \langle x \rangle$ with $f\phi_2 \neq 0$. Together, ϕ_1 and ϕ_2 extend to an epimorphism $\phi : C_p \wr A_n \to C_p \wr C_\infty$ with $f \notin \ker \phi$.

Write D for the intersection of the kernels of all epimorphisms $C_p \wr A_n \to C_p \wr C_\infty$. The previous paragraph shows that D has trivial intersection with the base group; as the base group is self-centralizing this implies that $D = 1$, which means that $C_p \wr A_n$ is residually $C_p \wr C_\infty$ as claimed.

Finally, for each $n \in \mathbb{N}$ we can embed $C_p^{(n)} \wr A_n$ in the direct product $(C_p \wr A_n)^{(n)}$, by the obvious mapping

$$(u_1, \ldots, u_n) \cdot a \mapsto (u_1 \cdot a, \ldots, u_n \cdot a)$$

(u_i in the base group, $a \in A_n$).

Now suppose that the word w satisfies $w(C_p \wr C_{p^n}) = 1$ for all $n \in S$. The first paragraph shows that then $w(C_p \wr C_\infty) = 1$; then $w(C_p \wr A_n) = 1$ for each n, by the second paragraph, and so $w(C_p^{(n)} \wr A_n) = 1$ for each n by the last paragraph. But $C_p^{(n)}$ and A_n are the free groups in $\mathfrak{A}_p$ and $\mathfrak{A}$ respectively; the result now follows by Lemma 4.4.4. $\blacksquare$

Corollary 4.4.6 *Let p be a prime and let $F = F_k$ where $k \in \mathbb{N}$. Then $F/F''(F')^p$ is residually $\{C_p \wr C_{p^n} \mid n \in S\}$ for any infinite set S of positive integers. Also the following are equivalent for $w \in F$:*

(a) *$w(C_p \wr C_{p^n}) = 1$ for infinitely many n;*

(b) *$w(C_p \wr C_\infty) = 1$;*

(c) *$w \in F''(F')^p$.*

The first claim follows from Propositions 4.4.5 and 4.4.3, and it implies that (c) follows from (a). It is obvious that (c) $\implies$ (b) $\implies$ (a).

More on wreath products and group varieties can be found in [N].

4.5 Words of infinite width in pro-p groups

Here we establish

Theorem 4.5.1 [J1] *Let w be a word, considered as an element of $F = F_k$. If $1 \neq w \in F''(F')^p$ then w has infinite width in the free pro-p group on two generators.*

Recall that $w \in F = F_k$ is said to be a J(p) word if $w \notin F''(F')^p$, and a J(π) word if it is a J(p) word for every p in the set π of primes. Combining Theorem 4.5.1 with Theorem 4.3.6 we obtain the promised characterization of words that are uniformly elliptic in the class $\mathfrak{N}(\pi)$ of all finite nilpotent π-groups:

Theorem 4.5.2 *Let π be a non-empty set of primes and let w be a non-trivial word. Then the following are equivalent:*

(a) *for each $p \in \pi$, w has bounded width in 2-generator finite p-groups;*

(b) *w is uniformly elliptic in $\mathfrak{N}(\pi)$;*

(c) *w is a J(π) word.*

Proof. If (a) holds then for each $p \in \pi$, w has finite width in the 2-generator free pro-p group, by Proposition 4.1.2; this implies (c) in view of Theorem 4.5.1. That (c) implies (b) is the content of Theorem 4.3.6; and (b) trivially implies (a). ∎

The equivalence of (a) and (b) is doubly remarkable: it shows that if the width of w in *two-generator p-groups* is bounded for each $p \in \pi$, then for each natural number m it is *uniformly* bounded in all m-generator p-groups, over all $p \in \pi$.

The proof of Theorem 4.5.1 needs some preparation. In Section 3.2 we defined the *Hausdorff dimension* for subsets of a finite group. This is extended to subsets of a finitely generated pro-p group L as follows. Let $(L_n)_{n \geq 1}$ denote the dimension subgroup series ('Zassenhaus filtration') of the pro-p group L; this is the fastest descending chain of open normal subgroups in L such that

$$[L_n, L] \leq L_{n+1}, \quad L_n^p \leq L_{pn}$$

for each n; see [DDMS], Chapter 11. Write $\pi_n : L \to L/L_n$ for the quotient map. Then the *Hausdorff dimension* of a subset S of L is

$$\text{hdim}(S) = \liminf_{n \to \infty} \text{hdim}_{L\pi_n}(S\pi_n) = \liminf_{n \to \infty} \frac{\log |S\pi_n|}{\log |L\pi_n|}.$$

Since $|(S^{*t})\pi_n| \leq |S\pi_n|^t$ for each n, we see that

$$\mathrm{hdim}(S^{*t}) \leq t \cdot \mathrm{hdim}(S)$$

for each $t \in \mathbb{N}$.

Now we quote the following proposition, which depends on the theory of free Lie algebras; the proof, sketched in Lemma 4.3 of [J1], is based on results in [G3], [NSS] and [P].

Proposition 4.5.3 *Let L be the free pro-p group on $d \geq 2$ generators, and let $N \neq 1$ be a closed normal subgroup of L. Say*

$$|L : L_{n+1}| = p^{b_n}, \quad |NL_{n+1} : L_{n+1}| = p^{h_n}.$$

Then

$$b_n \sim h_n \sim \frac{d^{n+1}}{(d-1)n} \quad \text{as } n \to \infty.$$

(Here, $f(n) \sim g(n)$ means that $f(n)/g(n) \to 1$ as $n \to \infty$.)

Corollary 4.5.4 *Let L be the free pro-p group on $d \geq 2$ generators.*

  (i) *If $N \neq 1$ is a closed normal subgroup of L then $\mathrm{hdim}(N) = 1$.*

  (ii) *Let v be the word $[x,y]z^p$. Then $\mathrm{hdim}(L_v) \leq 3/d$.*

Proof. (i) Taking logarithms to base p we have

$$\frac{\log |N\pi_n|}{\log |L\pi_n|} = \frac{h_{n-1}}{b_{n-1}} \to 1$$

as $n \to \infty$.

  (ii) Since L_n is central in L modulo L_{n+1} and $L_n^p \leq L_{pn} \leq L_{n+1}$, we see that L_n is marginal for v modulo L_{n+1}. Therefore v takes at most $|L/L_n|^3 = p^{3b_n - 1}$ values in L/L_{n+1}. So putting $S = L_v$ we have

$$\frac{\log |S\pi_{n+1}|}{\log |L\pi_{n+1}|} \leq \frac{3b_{n-1}}{b_n} \sim \frac{3n}{d(n-1)} \sim \frac{3}{d} \quad \text{as } n \to \infty.$$

(*Exercise.* Show that $v(x,y,z)^{-1} = v(y,x,z')$ for some z', so L_v really consists of positive values of v.) ∎

Note that (ii) is quite delicate:

Exercise 4.5.1. Keeping the above notation, prove that $\mathrm{hdim}(L_v) \geq 1/(d+1)$. [*Hint:* recall Exercise 4.2.3.]

Next, let H denote the free pro-p group on two generators and put $E = H'$. Then E is closed in H, by Theorem 4.1.5. Every closed subgroup of a free pro-p

group is free ([W4], Theorem 5.4.6); so E is a free pro-p group. *Warning*: the definition given above for the free pro-$\mathcal{C}$ group on a set X only applies when X is *finite*; for the general case, see §5.1 of [W4]. In this case, E is not finitely generated: to see this, note that for each n, H maps onto $P_n := C_p \wr C_{p^n}$, so E maps onto P'_n, which is elementary abelian of rank $p^n - 1$ (*Exercise!*).

Now let $1 \neq b \in E$ and let $n \in \mathbb{N}$. Then there exists $K \lhd_o E$ such that $b \notin K$ and $d(E/K) \geq n$. Put $Q = E/K$, let L be the free pro-p group on $d = d(Q)$ generators and choose an epimorphism $\phi : L \to Q$. Since E is free, we can lift the quotient map $\pi : E \to Q$ to a homomorphism $\psi : E \to L$ so that $\psi \circ \phi = \pi$. Then $\ker \psi \leq \ker \pi = K$ so $b\psi \neq 1$. Moreover, ψ maps E surjectively onto L; to see this, note that ϕ induces an isomorphism $L/\Phi(L) \to Q/\Phi(Q)$, and a glance at the commutative diagram

$$
\begin{array}{ccccc}
E & \overset{\pi}{\twoheadrightarrow} & Q & \twoheadrightarrow & Q/\Phi(Q) \\
\psi \searrow & & \uparrow \phi & & \uparrow \simeq \\
& & L & \twoheadrightarrow & L/\Phi(L)
\end{array}
$$

shows that $E\psi \cdot \Phi(L) = L$, which implies that $L = E\psi$.

We have established

Lemma 4.5.5 *Let $1 \neq b \in E$ and let $n \in \mathbb{N}$. Then for some $d \geq n$ there exists an epimorphism $\psi : E \to L_d$ with $b\psi \neq 1$, where L_d denotes the free pro-p group on d generators.*

To prove Theorem 4.5.1, let $1 \neq w \in F''(F')^p$ and suppose that w has finite width m in H, the free pro-p group on two generators; we aim to derive a contradiction. Write $E = H'$ and for each d let L_d denote the free pro-p group on d generators.

The hypotheses imply that $w(H) = W$ is closed and that $1 \neq W \leq E'E^p$. In view of Lemma 4.5.5, we may choose a large integer d so that there exists an epimorphism $\psi : E \to L = L_d$ with $W\psi \neq 1$.

Since $w \in F''(F')^p = v(F')$, we have

$$
w \in (F'_v)^{*t}
$$

for some t. It follows that

$$
W = H_w^{*m} \subseteq E_v^{*tm}
$$

and hence that

$$
W\psi \subseteq L_v^{*tm}.
$$

Using Corollary 4.5.4 we deduce:

$$
1 = \mathrm{hdim}(W\psi) \leq \mathrm{hdim}(L_v^{*tm})
$$

$$
\leq tm \cdot \mathrm{hdim}(L_v) \leq \frac{3tm}{d}.
$$

We get a contradiction by choosing $d > 3tm$.

This completes the proof of Theorem 4.5.1. Note that the basic idea is essentially the one we used in Section 3.2: one has to establish the existence of finite p-groups in which a central elementary abelian subgroup has large order compared with its index. This is implicitly the content of the easier half of Proposition 4.5.3.

4.6   Finite simple groups

So far, as regards the behaviour of word mappings in finite nilpotent groups, the results we have seen more or less mirror what happens in nilpotent groups in general. The really startling implications of finiteness, hinted at in the introduction to §4.1, become apparent when we turn to the world of simple groups.

The complete classification of the finite simple groups (CFSG) was announced around 1980, though the final component of the proof only appeared in 2004. A revised version of the proof is being published in a series of volumes beginning with [GLS]. The extreme length and technical complexity of the proof mean that many group theorists are unlikely to master the whole thing; but the *result* is by an order of magnitude the single most important fact about groups. It articulates a fundamental, very deep feature of mathematical reality: the ultimate building blocks of (finite) symmetry are these (a list of explicitly described groups); there are no others.

In practice, this means that results obtained by the detailed examination of specific types of group, such as alternating groups and certain matrix groups over finite fields, may lead to conclusions about all finite groups. A striking example of this phenomenon is Theorem 4.2.2 – on the face of it a result about compact topological groups that makes no mention of finite simple groups, yet (as far as we can tell) one that reflects in a subtle way the 'deep feature of mathematical reality' mentioned above.

In this section I can only report on some more or less recent results about verbal width in simple groups. Even taking CFSG as given, their proofs are long and involve a wide range of difficult mathematics, encompassing character theory, combinatorics, algebraic geometry and analytic number theory.

If w is a word then for any simple group G either $w(G) = 1$ or $w(G) = G$. The second case is 'generic', in the sense that most simple groups do not satisfy any given identity; indeed, it is a consequence of CFSG that for any word $w \neq 1$ only a *finite number* of finite simple groups G satisfy $w(G) = 1$ [J3]. Thus usually one has $G = w(G)$. Underlying the results we are about to discuss seems to be the surprising phenomenon that, generally speaking, subsets like G_w already make up a large proportion of the group G.

The following extraordinary result is due to Aner Shalev:

Theorem 4.6.1 [S10] *Every word has width* 3 *in every sufficiently large finite simple group.*

Here, 'sufficiently large' depends on the word, but of course the result implies

Corollary 4.6.2 *Every word is uniformly elliptic in finite simple groups.*

This was first established in [LS2]. (Every finite simple group can be generated by two elements [AG], so 'uniformly elliptic' here actually means 'of uniformly bounded width'.)

The theorem is far from the end of the story; in [LS1] Larsen and Shalev make the

Conjecture Let w_1 and w_2 be non-trivial words. Then

$$G_{w_1} \cdot G_{w_2} = G$$

for every sufficiently large finite simple group G.

In the same paper, they prove the analogous conjecture with three words instead of two, and the conjecture as stated for alternating groups and for groups of Lie type of bounded Lie rank (for a 'global' result lying behind this, see Corollary 5.1.5).

There are also conjectures and results about specific words. Shalev proves in [S10] that for a certain constant N, *each of the words γ_c ($c \geq 2$) has width 2 in every finite simple group of order at least N.* A long-standing conjecture of Ore states that *in a finite simple group every element is a commutator*, i.e. that γ_2 has width one in every finite simple group. The proof of Ore's conjecture has just been announced by Liebeck, O'Brien, Shalev and Tiep, a spectacular achievement combining algebraic geometry, character theory and serious computation.

As is usual when applying CFSG, the proofs of these results divide into two cases: (a) alternating groups and (b) simple groups of Lie type. A different approach to case (b) is given in [NP], based on an interesting discovery due to Tim Gowers. This was generalized by Babai, Nikolov and Pyber as follows. In this result, $k(G)$ denotes *the minimal dimension of a non-trivial $\mathbb{R}$-linear representation of G.*

Theorem 4.6.3 [BNP] *Let $X_1, \ldots, X_t$ be subsets of a finite group G, where $t \geq 3$. If*

$$\prod_{i=1}^{t} |X_i| \geq \frac{|G|^t}{k(G)^{t-2}}$$

then $X_1 \cdot X_2 \cdot \ldots \cdot X_t = G$.

Note that this applies to *any* finite group; in fact, the proof is purely combinatorial, and is quite accessible in [BNP] (in particular, it makes no reference to CFSG, or indeed to any structural aspects of group theory). The group theory comes into play when we want to apply the result. In particular one needs a lower bound for $k(G)$; such lower bounds are known for the simple groups. If G is simple of Lie type over $\mathbb{F}_q$, of (untwisted) Lie rank r and dimension d, then

$$k(G) \geq cq^r,$$

where c is an absolute constant, while $|G| \sim q^d$.

For the application to Theorem 4.6.1, let w be a non-trivial word and set

$$X_1 = X_2 = X_3 = G_w.$$

Then to show that $G_w^{*3} = G$ it suffices to verify that

$$|G_w| \geq |G| / k(G)^{1/3}.$$

That this is the case when G is a simple group of Lie type and sufficiently large order is deduced in [NP] from results in [LS1].

Theorem 4.6.1 had several precursors; the next two (earlier and easier) results were in a sense the 'base case' for the general results to be discussed in the following section.

Theorem 4.6.4 ([W3], Proposition 2.4) *The word γ_2 is uniformly elliptic in finite simple groups.*

Theorem 4.6.5 ([MZ], [SW]) *For each $q \in \mathbb{N}$ the word x^q is uniformly elliptic in finite simple groups.*

The – relatively short, modulo CFSG – proofs of these results are not 'effective': they do not give explicit bounds for the width of the word in question. The reason for this is their reliance on an interesting model-theoretic argument, which in effect models an infinite family of groups of a given Lie type over finite fields by a single group of that type over an infinite field. In the more recent work of Shalev and Larsen, arguments of this kind are replaced by ones using algebraic geometry.

Quasisimple groups

Each finite perfect group G (i.e. one with $G = G'$) has a 'universal covering group' $\widetilde{G}$, the biggest perfect group satisfying $\widetilde{G}/Z(\widetilde{G}) \cong G$. The centre $Z(\widetilde{G}) = M(G)$ is the *Schur multiplier* of G. If G is simple then $\widetilde{G}$ is a *quasisimple* group: a perfect group that is simple modulo its centre.

Most of the simple groups have small Schur multipliers. In fact, for a finite simple group G we have

$$|M(G)| \leq 48$$

unless G is of the form $\mathrm{PSL}(n, q)$ or $\mathrm{PSU}(n, q)$, in which case $M(G)$ is cyclic of order $\gcd(n, q \pm 1)$.

Using this, one can extend many results about simple groups to quasisimple groups. Theorem 4.6.3 provides one crude but simple technique:

Corollary 4.6.6 *Let $N \lhd G$ be finite groups. Suppose that*

$$|N| \leq k(G)^{1-2/t}$$

where $t \geq 3$. If the word w has width m in G/N then w has width tm in G.

Proof. Take $X_1 = X_2 = \cdots = X_t = G_w^{*m}$ in Theorem 4.6.3. Then $NX_i = G$ for each i so

$$\prod_{i=1}^{t} |X_i| \geq |G/N|^t \geq \frac{|G|^t}{k(G)^{t-2}}.$$

∎

Corollary 4.6.7 *There is an absolute constant K such that if G is a quasisimple group with $|G/Z(G)| \geq K$ and w is a word having width m in $G/Z(G)$ then w has width $3m$ in G.*

Proof. We apply the preceding corollary with $N = Z(G)$ and $t = 3$. Since G is quasisimple, $k(G) = k(G/N)$ which tends to infinity with $|G/N|$, so

$$k(G)^{1/3} > 48 \geq |N|$$

if G/N is large enough and G/N is not of the form $\mathrm{PSL}(n,q)$ or $\mathrm{PSU}(n,q)$. In these last two cases, one has $k(G) = k(G/N) \geq \frac{1}{2}(q^{n-1} - 1)$ and $|N| \leq \max\{n, q+1\}$. This gives $k(G)^{1/3} > |N|$ if $n \geq 7$ or $q \geq 8$ (very crudely). ∎

More careful estimates given in [NP] show that in fact every word has width 3 in all sufficiently large quasisimple groups of type $\mathrm{SL}(n,q)$ or $\mathrm{SU}(n,q)$.

Let us call a finite group G *quasi-semisimple* if G is perfect and $G/Z(G)$ is a direct product of simple groups. Then G is a quotient of its universal cover $\widetilde{G}$, and $\widetilde{G}$ is a direct product of quasisimple groups. Combining the preceding corollary with Theorem 4.6.1 we deduce

Corollary 4.6.8 *Let w be a word. Then there exists $f = f(w)$ such that w has width f in every finite quasi-semisimple group.*

The general theorems discussed in the following section ultimately reduce to two results about quasisimple groups, which may be seen as generalizations of Theorems 4.6.4 and 4.6.5. The first one considers 'twisted commutators': if α and β are automorphisms of a group G, we write

$$T_{\alpha,\beta}(x,y) = x^{-1}y^{-1}x^\alpha y^\beta$$

for $x, y \in G$, and set $T_{\alpha,\beta}(G,G) = \{T_{\alpha,\beta}(x,y) \mid x, \ y \in G\}$.

Theorem 4.6.9 [NS2] *There is an absolute constant D such that if S is any finite quasisimple group and $\alpha_1, \beta_1, \ldots, \alpha_D, \beta_D$ are any $2D$ automorphisms of S then*

$$S = T_{\alpha_1,\beta_1}(S,S) \cdot T_{\alpha_2,\beta_2}(S,S) \cdot \ldots \cdot T_{\alpha_D,\beta_D}(S,S).$$

Taking $\alpha_i = \beta_i = 1$, this extends Theorem 4.6.4 to quasisimple groups.

Theorem 4.6.10 [NS2] *Let q be a natural number. There exist natural numbers $C = C(q)$ and $M = M(q)$ such that if S is a finite quasisimple group with $|S/Z(S)| > C$, $\beta_1, \ldots, \beta_M$ are any M automorphisms of S, and $q_1, \ldots, q_M$ are any M divisors of q, then there exist inner automorphisms $\alpha_1, \ldots, \alpha_M$ of S such that*

$$S = [S, (\alpha_1 \beta_1)^{q_1}] \cdot \ldots \cdot [S, (\alpha_M \beta_M)^{q_M}].$$

Taking $\beta_i = 1$ and $q_i = 1$ we again recover Theorem 4.6.4; taking $\beta_i = 1$ and $q_i = q$ for each i extends Theorem 4.6.5 to quasisimple groups, since each commutator $[s, \alpha^q]$ is a product of two qth powers.

It is worth remarking that Theorem 4.6.9 extends to quasi-semisimple groups ([NS1], Proposition 11.1; this is by no means a routine extension.)

4.7   The Nikolov-Segal theorems

In the paper [NS1] we established three results about words that are uniformly elliptic in finite groups. Although they look slightly different, their proofs have a common strategy; this originates in the proof that γ_2 is uniformly elliptic in nilpotent groups, and may be described as the method of successive approximation, or Newton's method, or Hensel's Lemma.

Let us recall Proposition 1.2.5. This says that if $G = G' \langle g_1, \ldots, g_m \rangle$ and $K \lhd G$ then

$$[K, G] = \prod_{i=1}^{m} [K, g_i] \cdot [K, {}_n G] \tag{4.7}$$

for every n. If G is nilpotent, we eventually kill off the 'error term' $[K, {}_n G]$ by taking n to be the nilpotency class. We are interested now in *finite* groups, but not necessarily nilpotent ones. When G is finite, the error term stabilizes at some n, i.e. $[K, {}_n G] = [K, {}_{n+1} G] = H$, say; thus $H = [H, G]$. If we show that

$$H = [H, G] \Longrightarrow H = \left(\prod_{i=1}^{m} [H, g_i] \right)^{*f} \tag{4.8}$$

for some f, it will follow that $[K, G] = \left(\prod_{i=1}^{m} [K, g_i] \right)^{*(f+1)}$. In the special case $K = G$ this implies that γ_2 has width $m(f + 1)$ in G. Of course, this is only useful if we know what f depends on; ideally, it should depend only on $\mathrm{d}(G)$.

We were unable to prove quite such a general result (see Problem 4.7.1 below). To state what we did prove, we need the following

Definition Let G be a finite group and H a normal subgroup. Then H is *acceptable* if

(i) $H = [H, G]$ and

(ii) whenever $Z < N \leq H$ are normal subgroups of G, the factor N/Z is not of the form S or $S \times S$ for a non-abelian simple group S.

For $q \in \mathbb{N}$ write $\beta(q)(x) = x^q$; this is the 'Burnside word' of exponent q. The main part of [NS1] consists of the proof of the following *Key Theorem*:

Theorem 4.7.1 *Let G be a finite group and H an acceptable normal subgroup of G.*

(A) *Suppose that $G = \langle g_1, \ldots, g_d \rangle$ and let $q \in \mathbb{N}$. Then*

$$H = \left(\prod_{i=1}^{d} [H, g_i] \right)^{*f_1(d,q)} \cdot H_{\beta(q)}^{*f_2(q)}.$$

(B) *Suppose that $G = \langle g_1, \ldots, g_d \rangle$. Then*

$$H = \left(\prod_{i=1}^{d} [H, g_i] \right)^{*f_3(d)} \cdot H_{\gamma_2}^{*D}.$$

(C) *Suppose that $\mathrm{d}(G) = d$ and that $\mathrm{Alt}(s)$ is not involved in G. If $G = H\langle g_1, \ldots, g_r \rangle$ then*

$$H = \left(\prod_{i=1}^{r} [H, g_i] \right)^{*f_4(d,s)}.$$

Here, the functions $f_1, \ldots, f_4$ depend only on the displayed arguments, and D is an absolute constant (given in Theorem 4.6.9); moreover $f_3(d) = 6d^2 + O(d)$.

The proof is too long to reproduce here; an outline of the basic strategy is given at the end of this section.

Problem 4.7.1 Can $f_4(d, s)$ be made independent of s in Theorem 4.7.1(C)? (Note that if the answer is 'yes', then parts (A) and (B) are redundant.)

Exercise 4.7.1. Why is a group of the form $G = H \times \langle g_1, \ldots, g_r \rangle$ not a counterexample to Theorem 4.7.1(C)? [This partly motivates our peculiar definition of 'acceptable'.]

Theorem 4.7.1 has implications for the width of commutators, of Burnside words, and of locally finite words.

1. Commutators

If G is a finite d-generator group, then every element of G' is a product of about $12d^3$ commutators. This is one of my favourite theorems, which is much deeper than it looks. It follows from

Theorem 4.7.2 *Let G be a finite d-generator group and let $K \triangleleft G$. Then*

$$[K, G] = \mathcal{X}^{*f(d)} \tag{4.9}$$

where $\mathcal{X} = \{[x, g] \mid x \in K, \ g \in G\}$ and $f(d) = 12d^3 + O(d^2)$.

Copying the proof of Proposition 4.1.2, (a) $\Longrightarrow$ (c), one deduces

Corollary 4.7.3 *Let G be a finitely generated profinite group and K a closed normal subgroup of G. Then $[K, G]$ is closed in G.*

In particular, taking $K = G, \ \gamma_2(G), \ldots, \gamma_{c-1}(G)$ in turn we see that $\gamma_c(G)$ is closed in G for any finitely generated profinite group G and each $c \geq 2$. With Proposition 4.1.3 this gives

Corollary 4.7.4 *Each lower-central word γ_c is uniformly elliptic in finite groups.*

This line of argument shows the advantage of using profinite language when discussing uniform properties of finite groups. The disadvantage is a certain loss of precision; to get explicit bounds we have to work a bit harder:

Exercise 4.7.2. Assuming Theorem 4.7.2, prove that the word γ_c has width $f(d)^{c-1}$ in every d-generator finite group. [*Hint*: argue by induction on c, and use identities of the type

$$[x_1^{\pm 1} \ldots x_m^{\pm 1}, g] = [x_1', g_1]^{\pm 1} \ldots [x_m', g_m]^{\pm 1}$$

where x_i' is conjugate to x_i and g_i is conjugate to g for each i.]

Proof of Theorem 4.7.2. Now G is a finite d-generator group and $K \triangleleft G$. We need to prove (4.9) where $\mathcal{X} = \{[x, g] \mid x \in K, \ g \in G\}$. If K is acceptable in G this is immediate from Theorem 4.7.1(B); for the general case, we have to replace K by a slightly smaller subgroup that is acceptable, and work our way down.

Let $\mathcal{T}$ denote the class of all non-abelian finite simple groups; for a group H let $\mathcal{N}_{\mathcal{T}}(H)$ denote the set of all $Y \triangleleft H$ such that $H/Y \in \mathcal{T}$.

We define normal subgroups $H_1 \geq H_2 \geq H_3 \geq H_4$ of G as follows:

$$H_1 = [K, _n G] = [H_1, G] \tag{4.10}$$

where n is chosen large enough so that $[K, _n G] = [K, _{n+1} G]$;

$$H_2 = K_{\mathrm{sol}} = [H_2, H_2]$$

is the soluble residual of K, that is, the smallest normal subgroup S of K such that K/S is soluble;

$$H_3 = \bigcap \mathcal{N}_{\mathcal{T}}(H_2),$$

so H_2/H_3 is the biggest semisimple quotient of H_2; and

$$H_4 = [H_3, H_2].$$

Now H_4 is the desired acceptable normal subgroup of G. Indeed, $H_4 = [H_4, H_2]$ by the Three-subgroup Lemma, since $H_2 = [H_2, H_2]$, so $H_4 = [H_4, G]$. 'Acceptability condition' (ii) is satisfied because the outer automorphism groups of S and of $S \times S$ are soluble for every non-abelian simple group S ('Schreier's Conjecture', see [GLS]); so if $Z < N \leq H_4$ are normal subgroups of G and N/Z is of the form S or $S \times S$ then $H_2 = NC_{H_2}(N/Z)$ since H_2 is perfect; it follows that $H_2/C_{H_2}(N/Z) \cong N/Z$ and hence that $C_{H_2}(N/Z) \geq H_3 \geq N$, which is impossible since N/Z is non-abelian.

Applying Theorem 4.7.1(B) we infer that

$$H_4 \subseteq \mathcal{X}^{*(df_3(d)+D)}.$$

Next, we observe that H_2/H_4 is perfect and semisimple modulo its centre; this implies that its universal covering group U is a direct product of quasisimple groups. Now Theorem 4.6.9 (applied to ordinary commutators) shows that $U = U_{\gamma_2}^{*D}$; it follows that

$$H_2 = H_4 \cdot (H_2)_{\gamma_2}^{*D} \subseteq H_4 \cdot \mathcal{X}^{*D}.$$

As H_1/H_2 is an acceptable normal subgroup of G/H_2, we may apply Theorem 4.7.1(B) again to get

$$H_1 \subseteq H_2 \cdot \mathcal{X}^{*(df_3(d)+D)}.$$

Finally, (4.7) shows that

$$[K, G] \subseteq H_1 \cdot \mathcal{X}^{*d}.$$

Putting it all together we see that (4.9) holds with

$$\begin{aligned} f(d) &= d + (df_3(d) + D) + D + (df_3(d) + D) \\ &= (2f_3(d) + 1)d + 3D \\ &= 12d^3 + O(d^2) \end{aligned}$$

since $f_3(d) = 6d^2 + O(d)$.

2. Powers

A profinite group G is said to be *universal* if every finite group occurs as an open section of G; that is, if for every finite group H there exist open subgroups $B \lhd A$ of G such that $A/B \cong H$. We say in this case that H is *involved in G*. It is clear that G is non-universal if and only if the invariant

$$\begin{aligned} \alpha(G) &:= \sup \{n \in \mathbb{N} \mid \mathrm{Alt}(n) \text{ is involved in } G\} \\ &= \sup_{Q \in \mathfrak{F}(G)} \alpha(Q). \end{aligned}$$

is finite.

Theorem 4.7.5 *Let $q, s \in \mathbb{N}$. The Burnside word $\beta(q)$ is uniformly elliptic in the class of all finite groups G with $\alpha(G) \leq s$.*

With Proposition 4.1.2 this gives

Corollary 4.7.6 *If G is a finitely generated non-universal profinite group then the subgroup $G^q = \beta(q)(G)$ is closed (and hence open) in G for each $q \in \mathbb{N}$.*

(The claim in parentheses follows from the positive solution of the Restricted Burnside Problem; see the proof of Corollary 4.7.11.)

Problem 4.7.2[1] Is the corollary true also if G is universal? Equivalently, is $\beta(q)$ uniformly elliptic in the class of all finite groups?

It will be clear from the proof below that the answer is 'yes' if Problem 4.7.1 has a positive solution.

Proof of Theorem 4.7.5. Let E be a finite d-generator group with $\alpha(E) \leq s$, put $\mathcal{X} = \{h^q \mid h \in E\}$ and let $G = \langle \mathcal{X} \rangle = E^q$. We have to show that

$$G = \mathcal{X}^{*f}$$

where f depends only on q, d and s. It will be convenient to use the set

$$\mathcal{Y} = \{[h, g] \mid g \in \mathcal{X}, \ h \in G\},$$

which is contained in $\mathcal{X}^{*2}$ (*Exercise!*)

Let $\mathcal{S}$ denote the class of non-abelian finite simple groups S with $S^q = 1$ or $|S| \leq C(q)$ (the constant appearing in Theorem 4.6.10), and let $\mathcal{T}$ denote the class of non-abelian finite simple groups not in $\mathcal{S}$; it follows from CFSG that $\mathcal{S}$ contains only finitely many groups up to isomorphism [J3]. Now put

$$K = \bigcap_{\theta \in \Theta} \ker \theta$$

where Θ is the set of all homomorphisms from G into $\mathrm{Aut}(S \times S)$ with $S \in \mathcal{S}$.

Next, we define normal subgroups $H_1 \geq H_2 \geq H_3 \geq H_4$ of G as in the proof of Theorem 4.7.2 above (using the new definition of the class $\mathcal{T}$).

As before we see that H_4 is an acceptable normal subgroup of G (because every non-abelian finite simple group is in either $\mathcal{S}$ or $\mathcal{T}$). Now by definition, G is generated by $\mathcal{X}$; say $G = \langle g_1, \dots, g_n \rangle$ with each $g_i \in \mathcal{X}$. Then Theorem 4.7.1(A) gives

$$H_4 \subseteq \mathcal{Y}^{*n f_1(n,q)} \cdot \mathcal{X}^{*f_2(q)}.$$

The trouble is that we don't know how to bound n; this is articulated in Problem 4.7.3. If we use Theorem 4.7.1(C) instead, we can get by without knowing too much about the generators of G; we only need generators modulo H_4, and we can build these up by working 'down from the top'. But there is a price to be paid: the resulting bounds will (as far as we know) depend on the parameter s.

---
[1] The answer is YES! See the Appendix.

The argument goes like this. Suppose we can show that

$$G = H_4 \cdot \mathcal{X}^{*m} \qquad (4.11)$$

where m depends only on s, d and q. The solution of the Restricted Burnside Problem ([Z1], [Z2]) shows that $|E : G| \leq b(d, q)$, a number depending only on d and q; it follows that $d(G) \leq \widetilde{d} = d \cdot b(d, q)$ and hence that

$$G = H_4 \langle g_1, \ldots, g_r \rangle$$

with each $g_i \in \mathcal{X}$ and $r \leq m \cdot \widetilde{d}$. Now we can apply Theorem 4.7.1(C) to infer that

$$H_4 = \left(\prod_{i=1}^{r} [H_4, g_i] \right)^{*f_4(\widetilde{d}, s)} \subseteq \mathcal{Y}^{*r f_4(\widetilde{d}, s)},$$

and conclude that

$$G \subseteq \mathcal{Y}^{*r f_4(\widetilde{d}, s)} \cdot \mathcal{X}^{*m} \subseteq \mathcal{X}^{*f}$$

where $f = 2r f_4(\widetilde{d}, s) + m$.

It remains to establish (4.11).

As before, the universal covering group of H_2/H_4 is a direct product of quasisimple groups, to which Theorem 4.6.10 may be applied (we have made sure that the simple composition factors all belong to the class $\mathcal{T}$). This gives

$$H_2 \subseteq H_4 \cdot \mathcal{Y}^{*M(q)}.$$

Again, H_1/H_2 is an acceptable normal subgroup of G/H_2; we need to find a suitable set of generators for G modulo H_1. This depends on the following lemma, whose proof is left as an *Exercise*:

Lemma 4.7.7 *Let $A = \langle X \rangle$ be an r-generator abelian group. Then*

$$A = \langle y_1^q, \ldots, y_r^q, X_0 \rangle$$

with $y_1, \ldots, y_r \in A$, $X_0 \subseteq X$ and $|X_0| \leq r\sigma(q)$, where $\sigma(q)$ denotes the number of distinct prime factors of q.

Taking $A = G/G'$, $X = G'\mathcal{X}$ and $r = \widetilde{d}$ we may infer that $G = G' \langle g_1, \ldots, g_\tau \rangle$ where $\tau = \widetilde{d}(1 + \sigma(q))$ and $g_i \in \mathcal{X}$ for each i. Then (4.7) gives

$$[K, G] = \prod_{i=1}^{\tau} [K, g_i] \cdot [K,_n G] \subseteq \mathcal{Y}^{*\tau} H_1. \qquad (4.12)$$

Next, let $N = N(\mathcal{S})$ denote the product of the orders $|\mathrm{Aut}(S \times S)|$ over all $S \in \mathcal{S}$ (counting each isomorphism type once). Then (very crudely)

$$|G : K| \leq N^{N^{\widetilde{d}}} = \rho,$$

say. It follows that $\mathrm{d}(K) \leq \rho \widetilde{d}$ and hence that

$$|G : [K,G]K^q| \leq \rho\, |K : [K,G]K^q| \leq \rho q^{\rho\widetilde{d}} = \lambda,$$

say. Thus by Lemma 1.1.2 we have

$$\begin{aligned}
G &= [K,G]K^q \cdot \mathcal{X}^{*\lambda} \\
&\subseteq [K,G] \cdot \mathcal{X}^{*(1+\lambda)} \subseteq H_1 \mathcal{X}^{*(2\tau+1+\lambda)},
\end{aligned}$$

in view of (4.12). As G is generated by $\widetilde{d}$ elements it follows that

$$G = H_1 \langle y_1, \ldots, y_\mu \rangle$$

where $\mu = \widetilde{d}(2\tau + 1 + \lambda)$ and $y_i \in \mathcal{X}$ for each i.

Now at last we may apply Theorem 4.7.1(C) to the pair H_1/H_2, G/H_2, to deduce that

$$H_1 \subseteq \mathcal{Y}^{*\nu} H_2$$

where $\nu = \mu \cdot f_4(\widetilde{d}, s)$. Putting the last few steps together we obtain (4.11) with

$$m = 2\nu + (2\tau + 1 + \lambda) + 2M(q).$$

3. Locally finite words

Recall that a word w is *d-locally finite* if every d-generator group G satisfying $w(G) = 1$ is finite. On the face of it, this property should be undetectable within the universe of finite groups (a witness to its failure must be an *infinite* d-generator group!). Relevant for us here are two of its consequences:

Lemma 4.7.8 *Let w be a d-locally finite word.*

(i) $|\mathbb{Z} : w(\mathbb{Z})|$ *is finite.*

(ii) *There exists $\delta \in \mathbb{N}$ such that if G is any finite d-generator group then there exists $Y \subseteq G_w$ with $|Y| \leq \delta$ such that $w(G) = \langle Y \rangle$.*

Proof. Write $F = F_d$. Then $|\mathbb{Z} : w(\mathbb{Z})| \leq |F/w(F)|$ is finite. Since $w(F)$ has finite index in F, it is finitely generated, and each generator is a product of finitely many w-values. Thus $w(F) = \langle Z \rangle$ for a certain finite subset Z of F_w. Set $\delta = |Z|$. ∎

Obviously (ii) holds even if the word 'finite' is removed; but here we want to concentrate on the finite case:

Definition The word w is *d-restricted* if the conclusions (i) and (ii) of Lemma 4.7.8 hold.

In general, $\delta(d, w)$ will denote the least integer δ satisfying (ii), if it exists; otherwise, set $\delta(d, w) = \infty$. Set

$$q(w) := |\mathbb{Z} : w(\mathbb{Z})|.$$

Note that if $q(w) = q$ is finite then

$$G_{\beta(q)} \subseteq G_w \tag{4.13}$$
$$G^q \leq w(G)$$

for every group G.

For any natural number q, the Burnside word $w = \beta(q)$ obviously satisfies (i). When q is large, $\beta(q)$ is not even 2-locally finite; but the positive solution to the Restricted Burnside Problem ([Z1], [Z2]) shows that there is a finite upper bound $b(d, q)$ for the order $|G/G^q|$ as G ranges over all *finite d-generator groups*. It follows in view of (4.13) that if w is any d-restricted word then

$$b(d, w) := \sup \{|G/w(G)| \mid G \text{ finite, } \mathrm{d}(G) \leq d\}$$
$$< \infty.$$

Exercise 4.7.3. Let w be a word. Suppose that $b(d, w)$ is finite, and that w has bounded width in all d-generator finite groups. Show that then w is d-restricted. [*Hint*: observe that $\mathrm{d}(w(G)) \leq 1 + (d-1)b(d, w)$ by Schreier's formula.]

Problem 4.7.3[2] For which q and d is the word $\beta(q)$ d-restricted?

With Exercise 4.7.3, the next theorem shows that the answer is the same as the answer to the question: *for which q and d is G^q closed in every d-generator profinite group G?*

Theorem 4.7.9 *Let $d \in \mathbb{N}$ and let w be a d-restricted word. Then there exists $f = f(w, d) \in \mathbb{N}$ such that w has width f in every d-generator finite group.*

Corollary 4.7.10 *Every locally finite word is uniformly elliptic in finite groups.*

(A word is *locally finite* if it is d-locally finite for every $d \in \mathbb{N}$.)

Corollary 4.7.11 *Let G be a d-generator profinite group and let w be a d-restricted word. Then $w(G)$ is open in G.*

Proof. If $w(G) \leq N \lhd_o G$ then $|G/N| \leq b(d, w)$. As the set $\mathcal{N}$ of all such N is closed under finite intersections, it follows that $\mathcal{N}$ is finite; so $\overline{w(G)} = \bigcap \mathcal{N}$ is open. The result follows since $w(G)$ is closed, by Theorem 4.7.9 and Proposition 4.1.2. $\blacksquare$

---

[2] The answer is 'all'. See the Appendix.

Remark.   The fact that $w(G)$ is *closed* in G follows from Theorem 4.7.9 independently of the solution to the Restricted Burnside Problem, which is not used in the proof; it is only needed for the corollary.

Proof of Theorem 4.7.9.   Let E be a d-generator finite group and w a d-restricted word, with $q(w) = q$ and $\delta(d, w) = \delta$. Put $\mathcal{X} = E_w$ and let $G = \langle \mathcal{X} \rangle = w(E)$. We will show that

$$G = \mathcal{X}^{*f} \tag{4.14}$$

where f depends only on q, δ and d.

From (4.13) we have

$$E_{\beta(q)} \subseteq \mathcal{X}.$$

Write

$$\mathcal{Y} = \{[h, g] \mid g \in \mathcal{X},\ h \in G\},\quad \mathcal{Y}_0 = \{[h, a^q] \mid a \in E,\ h \in G\};$$

so $\mathcal{Y}_0 \subseteq \mathcal{Y} \subseteq \mathcal{X}^{*2}$. Now define normal subgroups

$$K \ge H_1 \ge H_2 \ge H_3 \ge H_4$$

exactly as in the proof of Theorem 4.7.5, in the preceding subsection.

From the definition of δ there exist $g_1, \ldots, g_\delta \in \mathcal{X}$ such that $G = \langle g_1, \ldots, g_\delta \rangle$. As before, H_4 is an acceptable normal subgroup of G; Theorem 4.7.1(A) now shows that

$$H_4 = \left(\prod_{i=1}^{\delta}[H_4, g_i]\right)^{*f_1(\delta, q)} \cdot H_{4,\beta(q)}^{*f_2(q)}$$
$$\subseteq \mathcal{Y}^{*f_1(\delta, q)} \cdot \mathcal{X}^{*f_2(q)}.$$

As before, we have

$$H_2 \subseteq H_4 \cdot \mathcal{Y}_0^{*M(q)}.$$

Again, H_1/H_2 is an acceptable normal subgroup of G/H_2; applying Theorem 4.7.1(A) to this pair we obtain

$$H_1 = H_2 \cdot \left(\prod_{i=1}^{\delta}[H_1, g_i]\right)^{*f_1(\delta, q)} \cdot H_{1,\beta(q)}^{*f_2(q)}$$
$$\subseteq H_2 \cdot \mathcal{Y}^{*f_1(\delta, q)} \cdot \mathcal{X}^{*f_2(q)}.$$

Next, (4.7) gives

$$[K, G] = \prod_{i=1}^{\delta}[K, g_i] \cdot [K, {}_n G] \subseteq \mathcal{Y}^{*\delta} \cdot H_1.$$

Finally, $|G : K| \leq \rho = N^{N^\delta}$ where $N = N(\mathcal{S})$ is as defined before (note that N depends only on q); and as before we may infer that

$$|G : [K,G]K^q| \leq \rho q^{\rho\delta} = \lambda,$$

say. Now Lemma 1.1.2 gives

$$G = [K,G]K^q \cdot \mathcal{X}^{*\lambda} \subseteq [K,G] \cdot \mathcal{X}^{*(1+\lambda)}$$

since $[K,G]K^q = [K,G]K_{\beta(q)} \subseteq [K,G] \cdot \mathcal{X}$.
Together we obtain (4.14) with

$$f = (1+\lambda) + 2\delta + 4f_1(\delta,q) + 2f_2(q) + 2M(q).$$

■

Problem 4.7.4[3] Is every d-restricted word d-locally finite?

Presumably the answer is 'no', but this may not be easy to establish (a potential counterexample would be the word $\beta(q)$, where q is large enough to make the Burnside group F_d/F_d^q infinite, if at the same time $\beta(q)$ is d-restricted).

4. Further variations

The special case of Theorem 4.7.1(C) relating to *soluble* groups was first established in [S4], with an explicit bound $f_4(d) = 9(8d + 5)$ (the proof given in [NS1], when restricted to soluble groups, is slicker and gives a slightly better bound).

Exercise 4.7.4. Let $G = \langle g_1, \ldots, g_r \rangle$ be a finite soluble group and K a normal subgroup. Show that

$$[K,G] = \left(\prod_{i=1}^r [K,g_i] \right)^{*f}$$

where $f = r + f_4(r)$.
[*Hint*: define H_1 as in (4.10), then use (4.7) and Theorem 4.7.1(C).]

Exercise 4.7.5. Let $G = \overline{\langle g_1, \ldots, g_r \rangle}$ be a prosoluble group and K a closed normal subgroup. Show that

$$[K,G] = \left(\prod_{i=1}^r [K,g_i] \right)^{*f}$$

where f is as above.

---

[3]No, because every $\beta(q)$ is d-restricted. See the Appendix.

Exercise 4.7.6. Let G be a finitely generated prosoluble group and $N \lhd G$. Show that if G/N is perfect then $N = G$. [*Hint*: show that $G = \langle g_1, \ldots, g_r \rangle$ for some $g_1, \ldots, g_r \in N$, then apply the preceding exercise with $K = G$.]

Exercise 4.7.7. Let G be a finite d-generator soluble group and K a normal subgroup. Suppose that $G = [K, G] \langle g_1, \ldots, g_r \rangle$. Show that there exist $y_{ij} \in K$ ($i = 1, \ldots, r$, $j = 0, \ldots, d$) such that

$$[K, G] = \left(\prod_{i=1}^{r} \prod_{j=0}^{d} [K, g_i^{y_{ij}}] \right)^{*t}$$

where t depends only on r and d. [*Hint*: define H_1 as above and find a suitable set of generators for G modulo H_1.]

5. Virtually nilpotent quotients

We can now deliver the promised generalization of Corollary 4.1.6. Recall that for a group variety $\mathcal{V}$ and group G,

$$\mathcal{V}(G) = \langle w(G) \mid w \in W_{\mathcal{V}} \rangle \, ;$$

this is the smallest normal subgroup V of G such that $G/V \in \mathcal{V}$.

Theorem 4.7.12 *Let G be a finitely generated profinite group and let $\mathcal{V}$ be a group variety. If $G/\mathcal{V}(G)$ is virtually nilpotent then $\mathcal{V}(G)$ is closed in G.*

The *converse* of this theorem is discussed in the following section.

Proof. Write $V = \mathcal{V}(G)$. There exist $H \lhd G$ and $c \in \mathbb{N}$ such that G/H is finite and $\gamma_{c+1}(H) \leq V$. Now Theorem 4.2.2, which we deduced in §4.2 from Theorem 4.7.9, shows that H is open in G. Therefore H is again a finitely generated profinite group, so $\gamma_{c+1}(H)$ is closed in H by Corollary 4.7.3. Thus $\gamma_{c+1}(H)$ is closed in G.

Put $d = \mathrm{d}(G)$, and let D be the intersection of the kernels of all epimorphisms from the free group $F = F_d$ onto G/H. There are only finitely many such epimorphisms so D has finite index in F; it follows that $\widetilde{F} := F/\gamma_{c+1}(D)$ is a finitely generated virtually nilpotent group. Therefore every subgroup of $\widetilde{F}$ is finitely generated; hence there exist $w_1, \ldots, w_n \in W_{\mathcal{V}}$ such that

$$\mathcal{V}(\widetilde{F}) = \left\langle w_1(\widetilde{F}), \ldots, w_n(\widetilde{F}) \right\rangle$$
$$= v(\widetilde{F}) = \widetilde{F}_v^{*t}$$

where $v = w_1 * \cdots * w_n$, which has finite width t, say, in $\widetilde{F}$ by Theorem 5.1.

Now suppose that $\gamma_{c+1}(H) \leq N \lhd_o G$. There is an epimorphism $F \to G/N$ and this induces an epimorphism $\pi : \widetilde{F} \to G/N$. Then

$$VN/N = \mathcal{V}(G/N)$$
$$= \mathcal{V}(\widetilde{F})\pi = \widetilde{F}_v^{*t}\pi$$
$$= (G/N)_v^{*t} = G_v^{*t}N/N.$$

Hence

$$V \subseteq \bigcap_{\gamma_{c+1}(H) \leq N \lhd_o G} G_v^{*t}N = G_v^{*t}\gamma_{c+1}(H)$$

since $G_v^{*t}\gamma_{c+1}(H)$ is closed. But $G_v^{*t}\gamma_{c+1}(H) \leq V$, so we have equality and V is closed. $\blacksquare$

6. Proof of Theorem 4.7.1

To conclude this section, let me try to give a rough idea of how one proves Theorem 4.7.1. In each case, the theorem asserts the solvability of a certain type of equation:

$$w(y_1, \ldots, y_n, g_1, \ldots, g_r) = h \tag{4.15}$$

where $h \in H$ is the 'constant', the 'unknowns' $y_1, \ldots, y_n$ are to be found in H, and $g_1, \ldots, g_r$ are fixed parameters coming from G. In Case (A) for example we have $r = d$, $n = df_1(d, q) + f_2(q)$ and

$$w(y_1, \ldots, y_n, g_1, \ldots, g_r) = \prod_{i=1}^{f_1(d,q)} \prod_{j=1}^{d} [y_{ij}, g_j] \cdot \prod_{i=1}^{f_2(q)} z_i^q$$

where $y_1, \ldots, y_n$ are re-named $y_{11}, \ldots, y_{f_1(d,q),d}, z_1, \ldots, z_{f_2(q)}$.

What the integers $f_i(-)$ and the constant D in part (C) have to be emerges in the details of the proof; in principle they can be effectively determined. We have a good estimate only for f_3, namely

$$f_3(d) = 6d^2 + O(d).$$

The solvability of (4.15) is proved by induction on $|H|$. Choose $N \lhd G$ minimal subject to

$$1 < N = [N, G] \leq H.$$

It is easy to see that then N contains a unique normal subgroup Z of G maximal subject to $Z < N$. Thus N/Z is a minimal normal subgroup of G/Z; it is either elementary abelian or it is a direct product of (at least three, because H was acceptable) isomorphic simple groups.

Now put

$$K = \begin{cases} [Z, G] & \text{if} & [Z, G] > 1 \\ N & \text{if} & [Z, G] = 1 = N' \\ N' & \text{if} & [Z, G] = 1 < N' \end{cases}.$$

Applying the inductive hypothesis to G/K, we find $c \in K$ and $v_1, \ldots, v_n \in H$ such that

$$c \cdot w(v_1, \ldots, v_n, g_1, \ldots, g_r) = h.$$

To make the induction step, we seek to kill off the 'error term' c by finding $a_1, \ldots, a_n \in N$ such that

$$w(a_1 v_1, \ldots, a_n v_n, g_1, \ldots, g_r) = h.$$

Writing $\mathbf{a} = (a_1, \ldots, a_n)$, $\mathbf{v} = (v_1, \ldots, v_n)$, this is equivalent to

$$w'_{\mathbf{v}, \mathbf{g}}(\mathbf{a}, \mathbf{1}) = c. \qquad (4.16)$$

Thus the problem is to show that K is contained in the image of the generalized word mapping $N^{(n)} \to N$ defined by $(a_1, \ldots, a_n) \mapsto w'_{\mathbf{v}, \mathbf{g}}(\mathbf{a}, \mathbf{1})$.

Of course, the truth of this depends on the various initial hypotheses. To see what is involved, let's consider a special case, arising in the proof of (C):

$$w(y_1, \ldots, y_n, g_1, \ldots, g_r) = \prod_{i=1}^{f} \prod_{l=1}^{r} [y_{il}, g_l]$$

$$= \prod_{j=1}^{n} [y_j, g_j]$$

where $n = fr$, $y_{(i-1)r+l} = y_{il}$, $g_{(i-1)r+l} = g_i$. Assume that N is abelian and $Z = 1$, so $K = N$. For $\mathbf{a} \in N^{(n)}$ we have

$$w'_{\mathbf{v}, \mathbf{g}}(\mathbf{a}, \mathbf{1}) = \prod_{j=1}^{n} [a_j, g_j]^{\tau_j} = \phi(a_1, \ldots, a_n)$$

say, where

$$\tau_j = \tau_j(\mathbf{v}, \mathbf{g}) = v_j \prod_{l<j} [g_l, v_l].$$

Writing N additively and considering it as a G-module, we see that

$$\phi(N^{(n)}) = \sum_{j=1}^{n} N(g_j^{\tau_j} - 1).$$

Since $N = [N, G]$, to ensure that $\phi(N^{(n)}) = N$ it would suffice to know that

$$\left\langle g_j^{\tau_j(\mathbf{v}, \mathbf{g})} \mid j = 1, \ldots, n \right\rangle N = G.$$

Now we go back to the beginning and beef up the inductive hypothesis:

- equation (4.15) has a solution $\mathbf{y} = \mathbf{v}$ such that $g_1^{\tau_1(\mathbf{v}, \mathbf{g})}, \ldots, g_n^{\tau_n(\mathbf{v}, \mathbf{g})}$ generate G, where $g_{(i-1)r+l} = g_l$ for each i, l.

As we have seen, this now allows us to lift a solution $\mathbf{v}$ modulo N to a solution $\mathbf{a}.\mathbf{v}$ in G. But for the induction to work, this solution has to satisfy the extra condition

$$\left\langle g_1^{\tau_1(\mathbf{a}.\mathbf{v},\mathbf{g})}, \ldots, g_n^{\tau_n(\mathbf{a}.\mathbf{v},\mathbf{g})} \right\rangle = G. \tag{4.17}$$

The proof of the inductive step consists of two independent arguments.

(1) Since (in this case) ϕ is a surjective homomorphism $N^{(n)} \to N$,

$$\left| \phi^{-1}(c) \right| = |\ker \phi| \geq |N|^{n-1} = |N|^{fr-1}.$$

(2) The proof that the number of tuples $\mathbf{a} \in N^{(n)}$ for which (4.17) *fails* is strictly less than $|N|^{fr-1}$. For this step, we need f to be sufficiently large.

Together, (1) and (2) show that we can lift $\mathbf{v}$ to $\mathbf{a}.\mathbf{v}$ and achieve both requirements of the inductive hypothesis.

The proof of the theorem follows this pattern in each of the various cases. When N/Z is abelian but N is not, (1) reduces to the study of certain quadratic mappings over finite fields. When N/Z is non-abelian, it reduces to the study of equations in products of finite quasisimple groups; this requires some complicated combinatorial arguments, which ultimately reduce the problem to solving certain other equations in quasisimple groups: Theorems 4.6.9 and 4.6.10 are precisely tailored to ensure the solvability of these.

Part (2) involves some permutation group theory and representation theory. It is at this stage in the proof of (C) that the parameter s (the largest degree of an alternating section in the group) is used; whether it is essential, or merely an artefact of our proof, is not known.

4.8   Uniformly elliptic words

To save repetition, in this section I will say that a word w is *uniformly elliptic* if w is uniformly elliptic in the class of *all finite groups*: that is, the width of w in any finite group G is bounded by a function of $\mathrm{d}(G)$. In view of Propositions 4.1.2 and 4.1.3, this is equivalent to each of the properties

- w has finite width in every finitely generated profinite group;

- $w(G)$ is closed in G for every finitely generated profinite group G.

A necessary condition for this to hold is given in Theorem 4.5.2: w *must be a* J(p) *word for every prime* p; in this case let us say that w is a *J-word*. (Recall from §4.4 that w is a J(p) word if and only if $w(C_p \wr C_\infty) \neq 1$, or equivalently $w \notin F''(F')^p$, F being the free group on the variables in w.)

The same theorem shows that this condition is sufficient for w to be uniformly elliptic in finite *nilpotent* groups. Whether it is sufficient for w to be uniformly elliptic is not known; here we explore some partial results in that direction.

Definition Let $q \in \mathbb{N}$. Then $\mathfrak{R}_q$ denotes the class of all locally finite groups of exponent dividing q.

$\mathfrak{R}_q$ is called *the restricted Burnside variety of exponent* q. It is a deep fact that $\mathfrak{R}_q$ is indeed a variety:

Exercise 4.8.1. Show that this follows from the positive solution to the Restricted Burnside Problem: *for each* $d \in \mathbb{N}$ *there is a finite upper bound for the orders of all finite* d-*generator groups* G *satisfying* $G^q = 1$. (The proof of this was completed in 1991 by Zelmanov [Z1], [Z2].) [*Hint:* for each $k \in \mathbb{N}$ let $W(k)$ be a set of words that generates the smallest normal subgroup N of finite index in F_k such that $F_k^q \leq N$, and consider $\bigcup_{k \in \mathbb{N}} W(k)$.]

Theorem 4.8.1 *Let* G *be a finitely generated profinite group and let* $c, q \in \mathbb{N}$. *Then*

(i) $\mathfrak{R}_q(G)$ *is open in* G;

(ii) $\gamma_{c+1}(\mathfrak{R}_q(G))$ *is closed in* G;

(iii) $G/\gamma_{c+1}(\mathfrak{R}_q(G))$ *is virtually nilpotent.*

Proof. Say $\mathrm{d}(G) = d$, and put $K = \mathfrak{R}_q(F_d)$. Then K has finite index in F_d, so K is generated by finitely many elements $v_1, \ldots, v_m$ say. Considering these as words in $x_1, \ldots, x_d$, put

$$v = v_1 \circledast \cdots \circledast v_m.$$

Then $v(F_d) \geq K$ (in fact they are equal), so v is a d-locally finite word. It follows by Corollary 4.7.11 that $v(G)$ is open in G.

Now for any group H we have

$$v(H) = v_1(H) \ldots v_m(H) \leq \mathfrak{R}_q(H).$$

Hence in particular $v(G) \leq \mathfrak{R}_q(G)$, so $\mathfrak{R}_q(G)$ is open.

It follows that $\mathfrak{R}_q(G)$ is again a finitely generated profinite group, so $\gamma_{c+1}(\mathfrak{R}_q(G))$ is closed in $\mathfrak{R}_q(G)$ by the remark following Corollary 4.7.3. This implies (ii), and (iii) is clear since $G/\mathfrak{R}_q(G)$ is finite. ∎

Definition A group variety $\mathcal{V}$ is a *J-variety* if $C_p \wr C_\infty \notin \mathcal{V}$ for every prime p.

Thus $\mathcal{V}$ is a J-variety if and only if $W_\mathcal{V}$ contains a J(p)-word for every prime p; if $\mathcal{V}$ is the variety defined by a single word w, then $\mathcal{V}$ is a J-variety if and only if w is a J-word.

Let $\mathcal{Y}$ denote the smallest locally- and residually-closed class of groups that contains every virtually soluble group. (Thus $\mathcal{Y}$ contains in particular all groups that are residually virtually-soluble or locally virtually-soluble.)

Theorem 4.8.2 *Let* $\mathcal{V}$ *be a variety. Then the following are equivalent:*

(a) $\mathcal{V}$ *is a J-variety;*

(b) *there exist $c, q \in \mathbb{N}$ such that every finite group in $\mathcal{V}$ lies in $\mathfrak{N}_c \mathfrak{R}_q$;*

(c) *there exist $c, q \in \mathbb{N}$ such that every $\mathcal{Y}$-group in $\mathcal{V}$ lies in $\mathfrak{N}_c \mathfrak{R}_q$.*

Here, $\mathfrak{N}_c = \mathcal{V}_{\gamma_{c+1}}$ denotes the variety of groups that are nilpotent of class at most c. Trivially (c) implies (b), and (b) implies (a) because $C_p \wr C_{p^n}$ is not in $\mathfrak{N}_c \mathfrak{R}_q$ if n is sufficently large compared with c and q (cf. Exercise 4.3.1). The proof that (a) implies (c) is given in the following section.

This now yields a converse to Theorem 4.7.12:

Theorem 4.8.3 *Let $\mathcal{V}$ be a group variety. Suppose that $\mathcal{V}(E)$ is closed in E where E is the free profinite group on $d \geq 2$ generators. Then $\mathcal{V}$ is a J-variety and $G/\mathcal{V}(G)$ is virtually nilpotent for every d-generator profinite group G.*

Proof. According to Proposition 4.1.4 there exist a word $w \in W_{\mathcal{V}}$ and a natural number m such that $\mathcal{V}(Q) = w(Q) = Q_w^{*m}$ for every d-generator finite group Q. It follows by Theorem 4.5.2 that w is a J-word. Thus $\mathcal{V}$ is a J-variety, and Theorem 4.8.2 shows that every finite group in $\mathcal{V}$ lies in $\mathfrak{N}_c \mathfrak{R}_q$, where c and q depend only on $\mathcal{V}$.

Now put $V = \mathcal{V}(E)$, $R = \mathfrak{R}_q(E)$. Then for $N \triangleleft_o E$ we have $E/NV \in \mathfrak{N}_c \mathfrak{R}_q$, so $\gamma_{c+1}(NR) \leq NV$. Thus

$$V = \overline{V} = \bigcap_{N \triangleleft_o E} NV \geq \bigcap_{N \triangleleft_o E} \gamma_{c+1}(NR) \geq \gamma_{c+1}(\overline{R}).$$

As $\overline{R}$ is open in E it follows that E/V is virtually nilpotent. This gives the result since $G/\mathcal{V}(G)$ is a quotient of E/V if G is a d-generator profinite group. (For this argument we don't need to use the deeper fact that $\overline{R} = R$; that $\overline{R}$ is open follows from the fact that $|E : N| \leq |F_d/\mathfrak{R}_q(F_d)|$ whenever $R \leq N \triangleleft_o E$.) ∎

This shows that a word w is uniformly elliptic if and only if $G/w(G)$ is virtually nilpotent for every finitely generated profinite group G; but it doesn't tell us which words have this property. A necessary condition is that w be a J-word; to establish the sufficiency of this condition, one would have to replace the last step of the preceding proof with an argument to show that E/V being virtually nilpotent implies $V = \overline{V}$. This is (more or less) what Theorem 4.3.3 does when E is pronilpotent. Perhaps this could at least be generalized to the prosoluble case:

Exercise 4.8.2. Let $\mathcal{V}$ be a J-variety and let $V = \mathcal{V}(G)$ where G is a finitely generated profinite group. Show that $\overline{V}/V$ is a perfect group. [*Hint:* if this is false then there exists $K \triangleleft G$ with $V \leq K < \overline{V}$ such that G/K is virtually soluble; then use Theorems 4.8.2(c), 4.8.1 and 4.7.12 to derive a contradiction.]

Problem 4.8.1 Does there exist a prosoluble group H having a normal subgroup K such that H/K is perfect? (Note that H can't be finitely generated, by Exercise 4.7.6, and that H/K has no proper subgroups of finite index, by Exercise 4.2.6.)

If the answer is 'no', we may infer that $\mathcal{V}(G)$ is closed in G whenever $\mathcal{V}$ is a J-variety and G is a finitely generated prosoluble group; in particular, a word will be uniformly elliptic in finite soluble groups if and only if it is a J-word. The best I can do along these lines is

Theorem 4.8.4 *Let G be a finitely generated profinite group. If $\mathcal{V}$ is a J-variety and $G/\mathcal{V}(G) \in \mathcal{Y}$ then $\mathcal{V}(G)$ is closed in G.*

Indeed, Theorem 4.8.2 shows that $\gamma_{c+1}(\mathfrak{R}_q(G)) \leq \mathcal{V}(G)$ for some c and q. Thus $G/\mathcal{V}(G)$ is virtually nilpotent, by Theorem 4.8.1, and the result follows by Theorem 4.7.12.

If G is soluble then so is $G/\mathcal{V}(G)$; taking G to be a relatively free profinite group in the variety of soluble groups of a given derived length, we may infer

Corollary 4.8.5 *Let w be a word. Then each of the following implies the next:*

(a) *w is uniformly elliptic in finite soluble groups;*

(b) *w is a J-word;*

(c) *for each $l \in \mathbb{N}$, w is uniformly elliptic in finite soluble groups of derived length l.*

Instead of restricting the groups, we can try restricting the variety. Let us say that $\mathcal{V}$ is an $\mathfrak{X}$-*type variety* if it is generated by groups in the class $\mathfrak{X}$; when $\mathfrak{X}$ is subgroup-closed, this is equivalent to each of the conditions *(b)* $\mathcal{V}$ is generated by residually-$\mathfrak{X}$ groups, *(c)* every finitely generated free $\mathcal{V}$-group is residually-$\mathfrak{X}$ and *(d)* the free $\mathcal{V}$-group of countably infinite rank is residually-$\mathfrak{X}$ (Proposition 4.4.3).

Theorem 4.8.6 *Let $\mathcal{V}$ be a $\mathcal{Y}$-type variety. Then $\mathcal{V}(G)$ is closed in G for every finitely generated profinite group G if and only if $\mathcal{V}$ is a J-variety.*

Proof. 'Only if' follows from Theorem 4.8.3. Assume now that $\mathcal{V}$ is a J-variety. Then according to Theorem 4.8.2 there exist c and q such that every $\mathcal{Y}$-group in $\mathcal{V}$ lies in $\mathfrak{N}_c\mathfrak{R}_q$. Thus $\mathcal{V}$ is generated by groups in $\mathfrak{N}_c\mathfrak{R}_q$ and it follows that $\mathcal{V} \subseteq \mathfrak{N}_c\mathfrak{R}_q$. Now let G be a finitely generated profinite group. Then $\mathcal{V}(G) \supseteq \gamma_{c+1}(\mathfrak{R}_q(G))$. It follows by Theorem 4.8.1 that $G/\mathcal{V}(G)$ is virtually nilpotent, and then by Theorem 4.7.12 that $\mathcal{V}(G)$ is closed in G. ∎

Corollary 4.8.7 *Let w be a word such that $F_\infty/w(F_\infty)$ is in $\mathcal{Y}$. Then w is uniformly elliptic if and only if w is a J-word.*

This includes as special cases many of our previous results. It applies for example when w is a locally finite word or a repeated commutator γ_c; it does *not* apply to the Burnside words $\beta(q)$ when q is large.

Problem 4.8.2 Let $w = [x, y, \ldots, y]$ be an 'Engel word'. Is the relatively free group $F_\infty / w(F_\infty)$ residually finite?

This is a well-known open question. (Since finite Engel groups are nilpotent, residual finiteness is equivalent in this case to residual nilpotence.) If the answer is 'yes', we may conclude that Engel words are uniformly elliptic.

Exercise 4.8.3. Show that the class $\mathcal{Y}$ is subgroup-closed. [*Hint*: let $\mathcal{X}$ be the class of groups G such that every subgroup of G is in $\mathcal{Y}$. Show that $\mathcal{X}$ is locally- and residually-closed and contains every virtually soluble group.]

4.9 On J-varieties

Throughout this section, $\mathcal{V}$ will denote a J-variety. This means that $C_p \wr C_\infty \notin \mathcal{V}$ for each prime p. Thus for each p there is a word $w_p \in W_\mathcal{V}$ (a *J(p) word*) such that

$$\mathcal{V}(C_p \wr C_\infty) \geq w_p(C_p \wr C_\infty) \neq 1.$$

Lemma 4.9.1 *For each prime p there exists $n(p) \in \mathbb{N}$ with the following property: if G is a group with $w_p(G) = 1$ then $G^{n(p)}$ centralizes every elementary abelian normal p-section of G.*

Proof. Let $B = \langle b^W \rangle$ be the base group of $W = C_p \wr C_\infty = B \rtimes \langle x \rangle$. As $w_p(W)$ is a non-trivial normal subgroup of W, the index $|B : B \cap w_p(W)|$ is finite. Therefore $[B, x^{n(p)}] \leq w_p(W)$ for some $n(p) \geq 1$. Now let $\overline{G}$ be any quotient of G and $A \lhd \overline{G}$ an elementary abelian p-group. Let $a \in A$ and $y \in \overline{G}$. We have an epimorphism $\theta : W \to \langle a, y \rangle$ sending b to a and x to y, and then

$$[a, y^{n(p)}] = [b, x^{n(p)}]\theta \in w_p(W)\theta \leq w_p(\overline{G}) = 1.$$

The result follows. ∎

Lemma 4.9.2 *Let $G \in \mathcal{V}$ be finitely generated and metabelian. Then G has finite rank.*

Proof. Put $A = G'$. Then A is Noetherian as a module for G/A, with the conjugation action ([LR], §4.2). For each prime p, the finitely generated G/A-module A/A^p is centralized by $AG^{n(p)}$, by Lemma 4.9.1. Thus A/A^p is finite, since $G/AG^{n(p)}$ is finite. Now A has a free abelian subgroup E such that A/E is a π-group for some finite set of primes π ([LR], **4.3.3**). If $p \notin \pi$ then $E/E^p \cong A/A^p$ is finite; thus E has finite rank, and so A has finite torsion-free rank $r_0(A)$.

The torsion subgroup T of A is a π-group of finite exponent (because A is Noetherian for G/A), and T/T^p embeds in A/A^p for each prime p. Therefore

$$\mathrm{rk}(A) \leq r_0(A) + \max_{p \in \pi} \mathrm{rk}(A/A^p) < \infty.$$

The result follows, since $\mathrm{rk}(G/A) = \mathrm{d}(G/A) < \infty$. ∎

Now we fix the following **Notation**:

$$\Phi = F_2/\mathcal{V}(F_2),$$

the two-generator free group in $\mathcal{V}$;

$$r = \mathrm{rk}(\Phi/\Phi''); $$

note that r is finite by Lemma 4.9.2.

Lemma 4.9.3 *Let $G \in \mathcal{V}$ be finite and let M be a non-abelian minimal normal subgroup of G. Then $G^{r!}$ normalizes each minimal normal subgroup of M.*

Proof. M is a direct product of isomorphic simple groups, permuted by G. Let S be one of them, and let $x \in G$. Then

$$S^{\langle x \rangle} = S \times S^x \times \cdots \times S^{x^{l-1}}$$

where x^l normalizes S, inducing by conjugation an automorphism ξ of S. If ξ has a non-trivial fixed point in S, call it z and put $Z = \langle z \rangle$. If ξ acts fixed-point freely on S then ξ fixes some non-trivial Sylow subgroup P of S ([G2], Chapter 10, §1), and we put $Z = \mathrm{Z}(P)$ and let $1 \neq z \in Z$. In either case,

$$\langle Z, x \rangle = \left(Z \times Z^x \times \cdots \times Z^{x^{l-1}} \right) \langle x \rangle$$

contains the two-generator metabelian group $\langle z, x \rangle$ of rank at least l. Now $\langle z, x \rangle$ is an image of Φ/Φ'', so $l \leq r$; it follows that $x^{r!}$ normalizes S. ∎

Proposition 4.9.4 *There is a natural number t such that G^t is soluble for every finite group $G \in \mathcal{V}$.*

Proof. The variety $\mathcal{V}$ contains only finitely many non-abelian simple groups (this is established in [J3], modulo CFSG); let q denote the least common multiple of the exponents of their automorphism groups, and put $t = qr!$. Now let $G \in \mathcal{V}$ be finite and let M be a non-abelian chief factor of G. If $x \in G$ then $x^{r!}$ normalizes each simple component of M, so $x^{r!q}$ centralizes M. Thus G^t centralizes every non-abelian chief factor of G. It follows that G^t is soluble. ∎

Now we turn to soluble $\mathcal{V}$-groups.

Lemma 4.9.5 *If $G \in \mathcal{V}$ is finitely generated and residually nilpotent then G is virtually nilpotent.*

Proof. Note that G is residually (finite nilpotent), since every finitely generated nilpotent group is residually finite ([S3], Chapter 1). For each prime p, the pro-p

completion $\widehat{G}_p$ of G satisfies $w_p(\widehat{G}_p) = 1$, and hence is virtually nilpotent by Theorem 4.3.4. The result now follows by Proposition 4.3.7. ∎

The group Φ/Φ'' is a 2-generator metabelian group of finite rank; it is therefore virtually residually nilpotent (cf. [S1], which shows that every finitely generated metabelian group has this property; or [LR], **5.3.9**, **5.2.8** and **4.3.1**). It follows by Lemma 4.9.5 that Φ/Φ'' is virtually nilpotent. We fix integers k and m such that

$$\gamma_{k+1}(\Phi^m) \leq \Phi''.$$

Lemma 4.9.6 *Let $G \in \mathcal{V}$ and let A be an abelian normal subgroup of G. Then $[A, {}_{k+1}\, x^m] = 1$ for every $x \in G$.*

Proof. Let $a \in A$ and put $B = \langle a^{\langle x \rangle} \rangle$. Then $\langle a, x \rangle = B \langle x \rangle$ is an image of Φ/Φ''. So

$$[a, {}_{k+1}\, x^m] \in [B, x^m, {}_k\, x^m] \leq \gamma_{k+1}((B\langle x \rangle)^m) = 1.$$

The result follows. ∎

Lemma 4.9.7 *Suppose that $\Gamma \in \mathcal{V}$ is a soluble group and that $B^m = 1$ for every abelian normal subgroup B of Γ. Then $\Gamma^f = 1$ where $f = m^{k+3}$.*

Proof. By [S3], Proposition 3 of Chapter 2, Γ has a normal subgroup N such that $N' \leq Z(N) = C_\Gamma(N)$. Put $Z = Z(N)$. Then $Z^m = 1$, and this implies that $N^m \leq Z$. Let $x \in \Gamma$ and put $y = x^m$. Applying Lemma 4.9.6 twice we have $[N, {}_{k+1}\, y] \subseteq Z$ and $[Z, {}_{k+1}\, y] = 1$. Now 'stability group theory' gives $[N, y^{m^k}] \subseteq Z$ and $[z, y^{m^k}] = 1$, whence $[N, y^{m^{k+1}}] = 1$. Thus $y^{m^{k+1}} \in C_\Gamma(N) = Z$, whence $x^{m^{k+3}} = y^{m^{k+2}} = 1$. ∎

Proposition 4.9.8 *Let $G \in \mathcal{V}$ be finitely generated and soluble. Then G^f is nilpotent, where $f = m^{k+3}$ depends only on $\mathcal{V}$.*

Proof. Arguing by induction on the derived length of G, we may assume that G^f is abelian-by-nilpotent. Then G is virtually abelian-by-nilpotent, and hence satisfies *max-n* ([LR], §4.2). It follows that the Fitting subgroup N of G is nilpotent.

Suppose B/N is an abelian normal subgroup of G/N, and let $x \in B$. Then $N\langle x^m \rangle / N'$ is nilpotent by Lemma 4.9.6, and so $N\langle x^m \rangle$ is nilpotent (cf. [S3], Chapter 1, Corollary 12). As $N\langle x^m \rangle \lhd B \lhd G$ it follows that $N\langle x^m \rangle \leq \mathrm{Fit}(G) = N$. Thus $B^m \leq N$. Applying Lemma 4.9.7 we infer that $G^f \leq N$. ∎

Proposition 4.9.9 *There exist $s, h \in \mathbb{N}$ such that*

$$\gamma_{s+1}(G^h) = 1$$

for every two-generator virtually soluble group $G \in \mathcal{V}$.

Proof. Suppose that $Q = \Phi/N$ is a virtually soluble quotient of Φ. It follows by Proposition 4.9.4 that Q has a soluble normal subgroup $Q_1 \geq Q^t$. Then Q_1 is finitely generated, so Q_1^f is nilpotent by Proposition 4.9.8. Thus Q^{tf} is nilpotent. Also Q^{tf} has finite index in Q, and hence contains $\mathfrak{R}_{tf}(Q)$. Thus $\mathfrak{R}_{tf}(\Phi)N/N$ is nilpotent.

Now let D be the virtually-soluble residual of Φ, namely the intersection of all $N \lhd \Phi$ with Φ/N virtually soluble. We have seen that for each such N the quotient $\mathfrak{R}_{tf}(\Phi)N/N$ is nilpotent; it follows that $\mathfrak{R}_{tf}(\Phi)/D$ is residually nilpotent, and hence virtually nilpotent by Lemma 4.9.5 (note that $\mathfrak{R}_{tf}(\Phi)$ has finite index in Φ, so it is finitely generated and contains D). Therefore Φ/D is virtually nilpotent, so we have $\gamma_{s+1}(\Phi^h) \leq D$ for some s and h. The result follows, since every two-generator virtually soluble $\mathcal{V}$-group is an image of Φ/D. ∎

Our main result depends on a nice characterization of virtually nilpotent groups due to Burns, Macedońska and Medvedev. Let

$$u = \prod_{j=1}^{d_1} x_{i_j}, \quad v = \prod_{j=1}^{d_2} x_{l_j}$$

(where $i_1, \ldots, i_{d_1}, l_1, \ldots, l_{d_2} \in \{1, \ldots, k\}$) be positive words. A group G is said to satisfy the *positive law* $L = L(x_1, \ldots, x_k) : u \equiv v$ if $u(\mathbf{g}) = v(\mathbf{g})$ for every $\mathbf{g} \in G^{(k)}$; the *degree* of the law L is $\max\{d_1, d_2\}$, and L is *non-trivial* if its degree is positive. The following lemma is due to Mal'cev:

Lemma 4.9.10 [M1] *For each $n \in \mathbb{N}$ there is a positive law L_n of degree 2^n in two variables such that every nilpotent group of class n satisfies L_n.*

Proof. Put $u_1(x, y) = xy$. Then for $n \geq 1$ put $u_{n+1}(x, y) = u_n(x, y)u_n(y, x)$. Now let L_n be the law

$$u_n(x, y) = u_n(y, x).$$

Every abelian group satisfies L_1. Now let G be nilpotent of class $n + 1$ and suppose inductively that $G/Z(G)$ satisfies L_n. Then for $a, b \in G$ we have $u_n(a, b) = u_n(b, a)z$ with $z \in Z(G)$, so

$$u_{n+1}(a, b) = u_n(b, a)z \cdot u_n(b, a) = u_n(b, a) \cdot u_n(b, a)z = u_{n+1}(b, a).$$

Thus G satisfies L_{n+1}. ∎

If G^h is nilpotent of class s then of course G satisfies the positive law $L_s(x^h, y^h)$ of degree $h2^s$. As this depends only on the two-generator subgroups of G, Proposition 4.9.9 implies

Corollary 4.9.11 *Every virtually soluble $\mathcal{V}$-group satisfies the positive law $L_s(x^h, y^h)$.*

Now we quote

Theorem 4.9.12 ([BMM], Theorem A) *For each $n \in \mathbb{N}$ there exist $c(n)$ and $e(n) \in \mathbb{N}$ such that every virtually soluble group which satisfies a positive law of degree n lies in the variety $\mathfrak{N}_{c(n)}\mathfrak{R}_{e(n)}$.*

(Actually, the result proved by Burns, Macedońska and Medvedev is considerably more general than this.)

We are now ready for the main result. In the previous section we defined the class $\mathcal{Y}$ to be *the smallest locally- and residually-closed class of groups that contains every virtually soluble group.*

Theorem 4.9.13 *Let $\mathcal{V}$ be a J-variety. Then there exist $c, q \in \mathbb{N}$ such that*

$$\mathcal{V} \cap \mathcal{Y} \subseteq \mathfrak{N}_c \mathfrak{R}_q.$$

Proof. We take $c = c(h2^s)$ and $q = e(h2^s)$, where h and s are given in Corollary 4.9.11. Then Theorem 4.9.12 says that every virtually soluble $\mathcal{V}$-group lies in $\mathfrak{N}_c\mathfrak{R}_q$. Now let $\mathcal{V}^*$ denote the complement of $\mathcal{V}$ and consider the class of groups

$$\mathcal{X} = \mathfrak{N}_c\mathfrak{R}_q \cup \mathcal{V}^*.$$

Certainly $\mathcal{X}$ contains every virtually soluble group: for if G is virtually soluble and $G \notin \mathcal{V}^*$ then $G \in \mathcal{V}$, so $G \in \mathfrak{N}_c\mathfrak{R}_q$. On the other hand, since both $\mathcal{V}$ and $\mathfrak{N}_c\mathfrak{R}_q$ are varieties, $\mathcal{X}$ is both locally-closed and residually-closed; to see this, suppose that $(H_\alpha)_{\alpha \in A}$ is a family of subgroups or quotients of a group G with each H_α in $\mathcal{X}$. Then either $H_\alpha \in \mathcal{V}^*$ for some α, in which case $G \in \mathcal{V}^*$, or else $H_\alpha \in \mathfrak{N}_c\mathfrak{R}_q$ for every $\alpha \in A$, which implies that $G \in \mathfrak{N}_c\mathfrak{R}_q$ if G is locally or residually $\{H_\alpha \mid \alpha \in A\}$.

It follows that $\mathcal{X}$ contains $\mathcal{Y}$, which is precisely the claim of the theorem. ∎

Chapter 5

Algebraic and analytic groups

5.1 Algebraic groups

Here, *algebraic group* will mean a Zariski-closed subgroup of $\mathrm{SL}_n(K)$, for some $n \in \mathbb{N}$ and some algebraically closed field K. For background and terminology, see [B2], [H2], [PR] and [W1]. In this section, topological language refers to the Zariski topology.

The following theorem, due to Merzljakov, in a sense provides the philosophical background to all the ellipticity results concerning groups of Lie type; a sharper result specific to simple groups is stated below.

Theorem 5.1.1 [M2] *Every algebraic group is verbally elliptic.*

This depends on Chevalley's concept of *constructible sets*. Let $V = K^d$ be the affine space, with its Zariski topology. A subset of V is *constructible* if it is a finite union of sets of the form $C \cap U$ where C is closed and U is open. A *morphism* from V to $V_1 = K^l$ is a mapping defined by l polynomials.

The key result is

Proposition 5.1.2 *Let Y be a constructible subset of V, with closure $\overline{Y}$.*

(i) (See [W1], **14.9**) *The set Y contains a subset U that is open and dense in $\overline{Y}$.*

(ii) (Chevalley, see [H2], §4.4) *If $f : V \to V_1$ is a morphism then $f(Y)$ is a constructible subset of V_1.*

Now we can prove Theorem 5.1.1. Let w be a word in k variables. The algebraic group G is a closed subset of the affine space $V = \mathrm{M}_n(K) = K^{n^2}$. Recall that a subset S of V is *irreducible* if S is not the union of two proper (relatively) closed subsets of S. The connected component of 1 in G is a closed normal subgroup G^0; it has finite index in G, and its cosets are precisely the irreducible components of G.

Put $L = G^0$. According to a theorem of Platonov (see [W1], **10.10**), G has a finite subgroup H such that $G = LH$. Then

$$w(G) = w'_H(L)w(H),$$

and $w'_H(L) = \xi(L)$ where ξ is a certain generalized word function on L; if $H = \{h_1, \ldots, h_q\}$ we take

$$\xi = w'_{h_1} \divideontimes \cdots \divideontimes w'_{h_q}.$$

Now ξ maps $L^{(kq)}$ into L, and it is a *morphism*: indeed, $\xi(g_1, \ldots, g_{kq})$ is obtained from the matrices g_i by the operations of matrix multiplication, conjugation by certain fixed elements of H, and matrix inversion (which is a polynomial operation because $G \leq \mathrm{SL}_n(K)$).

Put $X = \mathrm{Im}\, \xi$ and $Y = X^{-1}$ (so $L_\xi = X \cup Y$), for $j \in \mathbb{N}$ write

$$P_j = (XY)^{*j}$$

and let $\overline{P_j}$ denote the closure of P_j. Observe that P_j is the image of the morphism $L^{(2kqj)} \to L$ that sends $(\mathbf{g}_1, \ldots, \mathbf{g}_{2j})$ to

$$\xi(\mathbf{g}_1)\xi(\mathbf{g}_2)^{-1} \ldots \xi(\mathbf{g}_{2j-1})\xi(\mathbf{g}_{2j})^{-1}$$

(each $\mathbf{g}_i \in L^{(kq)}$). Now $L^{(2kqj)}$ is a closed irreducible subset of $V^{(2kqj)}$; it follows easily that $\overline{P_j}$ is irreducible. Choose j so that $\overline{P_j}$ has maximal possible dimension (this is at most $\dim L \leq n^2 - 1$). If $l > j$ then $\overline{P_l} \supseteq \overline{P_j}$ is a pair of irreducible closed sets of equal dimension, which forces $\overline{P_l} = \overline{P_j}$. Hence

$$\xi(L) = \bigcup_{l=1}^\infty P_l \subseteq \overline{P_j} = T,$$

say. Thus $T = \overline{\xi(L)}$ is a closed subgroup of L ([W1], **5.9**).

It follows from Proposition 5.1.2 that P_j is constructible, and hence that P_j contains an open dense subset U of T. If $y \in T$ then left multiplication by y is a homeomorphism of T, so yU is a non-empty open subset of T. Therefore $yU \cap U \neq \varnothing$. Thus $y \in U \cdot U^{-1} \subseteq P_{2j}$.

We conclude that

$$\xi(L) = T = P_{2j} \subseteq L_\xi^{*4j} \subseteq G_w^{*8jq}.$$

Since w has width $q = |H|$ in H, it follows that

$$w(G) = \xi(L)w(H) = G_w^{*f}$$

where $f = q + 8jq$.

This completes the proof. (The essence of the argument is taken from [W1], Lemma 14.14.)

Since $\xi(L) = \overline{\xi(L)}$ is irreducible and has finite index at most $|w(H)| \leq |H|$ in $w(G)$, we have also established:

Corollary 5.1.3 (of proof) *Each verbal subgroup $w(G)$ is closed, and $\left|w(G) : w(G)^0\right|$ is bounded by a number depending only on G; $w(G)$ is connected if G is connected.*

This is as far as we get using elementary general arguments from algebraic geometry. By invoking the structure theory of algebraic groups, Borel established a much sharper result in a more restrictive context; here, $f_w : G^{(k)} \to G$ denotes the morphism given by evaluating the word w (so $G_w = f_w(G) \cup f_w(G)^{-1}$).

Theorem 5.1.4 ([B1], Theorem B) *Let G be a connected simple algebraic group and let w be a non-trivial word. Then $f_w(G)$ is Zariski-dense in G.*

Corollary 5.1.5 *If G is as above and u, v are non-trivial words then $G_u \cdot G_v = G$. Every word has width 2 in G.*

Proof. Now $\overline{G_u} = \overline{G_v} = G$; using Proposition 5.1.2 as in the preceding proof we deduce that G has dense open subsets U and V with $U \subseteq G_u$ and $V \subseteq G_v$. Now let $g \in G$. Then $g^{-1}U \cap V \neq \varnothing$, so $g \in UV^{-1} \subseteq G_u \cdot G_v$. The final claim is immediate. ∎

5.2   Adelic groups

We turn now to an interesting class of profinite groups. Let $G \leq \mathrm{SL}_n(\mathbb{C})$ be an algebraic group defined over $\mathbb{Q}$. For any integral domain R of characteristic zero, the solution-set in $\mathrm{SL}_n(R)$ of the equations defining G is denoted G_R. Thus if R is a subring of $\mathbb{C}$ we have

$$G_R = G \cap \mathrm{SL}_n(R),$$

showing that G_R is a subgroup of G. In particular, for each prime p we may fix an embedding of $\mathbb{Q}_p$ in $\mathbb{C}$ and get the groups $G_{\mathbb{Z}_p} \leq G_{\mathbb{Q}_p} \leq \mathrm{SL}_n(\mathbb{Q}_p)$. The group $G_{\mathbb{Q}_p}$ inherits a natural topology from $\mathbb{Q}_p$, making it a locally compact topological group; $G_{\mathbb{Z}_p}$ is a profinite group, with the subspace topology.

Let S be a finite set of primes, and let S' denote the set of all primes not in S. For each $q \in S'$ let

$$\rho_q : \prod_{p \in S'} G_{\mathbb{Q}_p} \to G_{\mathbb{Q}_q}$$

denote the projection to the q-component. The *S-adele group* of G is the group

$$G_{A_S} = \left\{ \mathbf{g} \in \prod_{p \in S'} G_{\mathbb{Q}_p} \mid \mathbf{g}\rho_q \in G_{\mathbb{Z}_q} \text{ for almost all } q \right\}$$

('almost all' means 'all but finitely many'). This becomes a locally compact topological group with the 'restricted product topology'; a base for the open sets in this topology is given by the sets

$$\prod_{p \in T'} G_{\mathbb{Z}_p} \times \prod_{p \in T \smallsetminus S} U_p \tag{5.1}$$

where U_p is an open subset of $G_{\mathbb{Q}_p}$ and T ranges over finite sets of primes containing S. (For all this see [PR], Chapter 5.) Inside G_{A_S} we have the open subgroup

$$G_\infty(S) := \prod_{p \in S'} G_{\mathbb{Z}_p};$$

note that the topology induced on this subgroup is the usual product topology, so $G_\infty(S)$ is a profinite group, and in particular it is compact. The following is more or less immediate from the definitions:

Lemma 5.2.1 *Let H be a closed subgroup of G_{A_S}. Then the following are equivalent:*

(a) *H is compact and open in G_{A_S};*

(b) *H is commensurable with $G_\infty(S)$.*

Here, (b) means that $H \cap G_\infty(S)$ has finite index both in H and in $G_\infty(S)$. For want of a better name, I will call a group H of this kind (where S is any finite set of primes) a *profinite form of G*.

Theorem 5.2.2 *Let G be a simply connected semisimple algebraic group defined over $\mathbb{Q}$, let H be a profinite form of G and w a non-trivial word. Then $w(H)$ is open in H.*

Corollary 5.2.3 *Every profinite form of G is verbally elliptic.*

The theorem looks very like Theorem 5.1.1; but it is much more delicate. The earlier result is essentially about the solvability of equations over an algebraically closed field, while here we are considering equations over the p-adic integers (for almost all primes p). The proof will accordingly depend on either some hard finite group theory ([LS2], [S10]), or alternatively some hard arithmetic geometry [LS1]. For some more remarks on this point see the end of this section.

The special case of Theorem 5.2.2 where $G = \mathrm{SL}_n$ and $H = G_\infty(\varnothing)$ was obtained by Andrei Jaikin (personal communication). A profinite group is said to be *adelic* if it is isomorphic to a closed subgroup of $\mathrm{SL}_{n,\infty}(\varnothing)$.

Problem 5.2.1 Is every adelic finitely generated profinite group verbally elliptic?

For some remarks on the status of this problem, see the Appendix.

Let w be a non-trivial word, S a finite set of primes, and H a compact open subgroup of G_{A_S}. Then H is a union of finitely many basic open sets like (5.1); hence there is a finite set of primes $S_1 \supseteq S$ such that the projection of H into $\prod_{p \in S_1'} G_{\mathbb{Q}_p}$ is equal to $G_\infty(S_1)$. If T is any finite set of primes containing S_1 we then have

$$H = \prod_{p \in T'} G_{\mathbb{Z}_p} \times H_T = G_\infty(T) \times H_T \qquad (5.2)$$

where $H_T = H \cap \prod_{p \in T \setminus S} G_{\mathbb{Q}_p}$.

For each prime p there is a natural homomorphism

$$\pi_p : G_{\mathbb{Z}_p} \to \mathrm{SL}_n(\mathbb{F}_p),$$

given by reducing each matrix entry modulo p. Note that

$$N_p := \ker \pi_p = G \cap \mathrm{SL}_n^1(\mathbb{Z}_p)$$

where

$$\mathrm{SL}_n^1(\mathbb{Z}_p) = \{g \in \mathrm{SL}_n(\mathbb{Z}_p) \mid g \equiv \mathbf{1}_n \pmod{p}\}.$$

It follows that N_p is a pro-p group of rank at most $n^2 - 1$ ($2n^2$ if $p = 2$; see [DDMS], Chapter 5; [LS3], Window 5).

Since G is defined over $\mathbb{Q}$, there is a finite set of primes T_1 such that the defining polynomial equations have coefficients in $\mathbb{Z}_{T_1} = \mathbb{Z}[1/p \mid p \in T_1]$. Then for $p \in T_1'$ we can reduce these equations modulo p; the resulting equations over $\mathbb{F}_p$ are satisfied by every matrix in $G_{\mathbb{Z}_p}\pi_p$. Making T_1 larger if necessary, we can ensure that these equations define an algebraic group $G^{(p)}$ over $\mathbb{F}_p$; that is, a Zariski-closed subgroup of $\mathrm{SL}_n(\overline{\mathbb{F}_p})$ where $\overline{\mathbb{F}_p}$ denotes the algebraic closure of $\mathbb{F}_p$. This is explained (in greater generality) in [PR], §3.3. The following further properties of this 'reduction map' are established there and in Window 9 of [LS3]:

Proposition 5.2.4 *There is a finite set of primes T_2 such that for each $p \in T_2'$,*

(i) *$G^{(p)}$ is a connected semisimple algebraic group;*

(ii) *$G^{(p)}_{\overline{\mathbb{F}_p}}$ is a product of fewer than $\dim G$ finite quasisimple groups, which are groups of Lie type over finite fields of characteristic p;*

(iii) *$G_{\mathbb{Z}_p}\pi_p = G^{(p)}_{\mathbb{F}_p}$;*

(iv) *N_p is contained in the Frattini subgroup $\Phi(G_{\mathbb{Z}_p})$ of $G_{\mathbb{Z}_p}$;*

(v) *$N_p = [N_p, G_{\mathbb{Z}_p}]$.*

(Part (iv) is Lemma 5 of [LS3], Window 9, where it is wrongly stated; the proof is correct, I hope. Claim (v) follows from the fact established there that $N_p/N_p'N_p^p$ is isomorphic as a $G^{(p)}_{\mathbb{F}_p}$-module to the Lie algebra of $G^{(p)}_{\mathbb{F}_p}$ with the adjoint action; this is a perfect module. It is assumed there that G is $\mathbb{Q}$-simple; in the semisimple case, G is a direct product of $\mathbb{Q}$-simple groups, and the results carry over directly for almost all primes.)

Let $\mathcal{S}(w)$ denote the set of finite simple groups S such that $w(S) = 1$; this is a finite set [J3], so by choosing T_2 sufficiently large we may ensure that the simple factors of $G^{(p)}_{\mathbb{F}_p}$ are not in $\mathcal{S}(w)$ when $p \in T_2'$.

Proposition 5.2.5 *Let Γ be a profinite group and $N \leq \Phi(\Gamma)$ an open normal subgroup. Suppose*

(a) *that Γ/N is the direct product of s quasisimple groups, that $N = [N,\Gamma]$ and that N is a pro-p group of rank r, for some prime p;*

(b) *that the simple factors of Γ/N are not in $\mathcal{S}(w)$.*

Then
$$\Gamma = \Gamma_w^{*f}$$
where $f = f(w,r,s)$ depends only on w, r and s.

Proof. Put $Z/N = \mathrm{Z}(\Gamma/N)$. Then $w(\Gamma)Z = \Gamma$; as Γ/N is perfect it follows that
$$\Gamma = \Gamma'N \le w(\Gamma)N \le w(\Gamma)\Phi(\Gamma)$$
and hence that $\Gamma = \overline{w(\Gamma)}$. Now Corollary 4.6.8 shows that
$$\Gamma = w(\Gamma)N = \Gamma_w^{*m} \cdot N$$

where m depends only on w (this is where the hard finite group theory comes in; in this case, as all the simple groups arising are Lie-type groups of bounded dimension, one can draw a similar conclusion from results in [LS1], proved using methods of arithmetic geometry).

As N is a pro-p group of finite rank, $\Phi(N)$ is open in N, and hence in Γ. Therefore $\Gamma = w(\Gamma)\Phi(N)$, and so $N/\Phi(N)$ is generated by the cosets of elements of the form abc^{-1} with $a,c \in \Gamma_w^{*m}$ and $b \in \Gamma_w$. As N has rank r it follows that
$$N = \overline{\langle g_1, \dots, g_r \rangle}$$

with each $g_i \in \Gamma_w^{*(2m+1)}$. Then by Corollary 4.3.2 we have
$$N' = \prod_{i=1}^{r}[N, g_i] \subseteq \Gamma_w^{*2r(2m+1)}.$$

If we show that $N \subseteq \Gamma_w^{*q}N'$, the result will follow with
$$f(w,r,s) = m + q + 2r(2m+1).$$

As N' is closed and normal in Γ, we may now replace Γ by Γ/N', and so reduce to the case where N is abelian.

Now Γ/Z can be generated by $2s$ elements, since each simple factor is a 2-generator group. Thus there exist $h_1, \dots, h_{2s} \in \Gamma_w^{*m}$ such that $\Gamma = \langle h_1, \dots, h_{2s} \rangle Z$, and, as before, the fact that Γ/N is perfect now implies that $\Gamma = \langle h_1, \dots, h_{2s} \rangle N$. As N is abelian it follows that
$$N = [N, \Gamma] = \prod_{i=1}^{2s}[N, h_i] \subseteq \Gamma_w^{*4sm},$$

and the result follows with $q = 2sm$. ∎

Together, the last two propositions show that, provided the finite set T of primes is suitably chosen,

$$G_\infty(T) = G_\infty(T)_w^{*f} = w(G_\infty(T)),$$

where f depends only on w and n; recall that both the rank of each N_p and the dimension of G are bounded by $n^2 - 1$. Thus

$$w(H) = w(G_\infty(T) \times H_T) = G_\infty(T) \times w(H_T);$$

so to complete the proof of Theorem 5.2.2, it will suffice to show that $w(H_T)$ is open in H_T.

Now H_T is commensurable with $\prod_{p \in \pi} G_{\mathbb{Z}_p}$ where $\pi = T \smallsetminus S$, so its subgroup $H \cap \prod_{p \in \pi} N_p$ is open both in H_T and in $\prod_{p \in \pi} N_p$. Hence there exist open subgroups $Q_p \le N_p$ such that

$$Q := \prod_{p \in \pi} Q_p \le_o H_T.$$

Since

$$w(H_T) \ge w(Q) = \prod_{p \in \pi} w(Q_p),$$

it will suffice to show that $w(Q_p)$ is open in Q_p, for each p.

Now fix a prime p. According to the structure theory of algebraic groups (see [PR], §2.1, [T], §3.1), there exist finitely many algebraic number fields k_i and for each i an absolutely simple algebraic group E_i defined over k_i such that G is $\mathbb{Q}$-isomorphic to

$$\prod_i \mathrm{R}_{k_i/\mathbb{Q}}(E_i)$$

('the restriction of scalars'). For each i we have $\mathbb{Q}_p \otimes k_i = \bigoplus_j K_{ij}$, where $K_{ij} = k_{\mathfrak{p}_{ij}}$ and the $\mathfrak{p}_{ij}$ are the primes of k_i dividing p. We can make the identification

$$G_{\mathbb{Q}_p} = \prod_{i,j} E_{i,K_{ij}},$$

and then the compact open subgroup Q_p of $G_{\mathbb{Q}_p}$ contains an open subgroup of the form $\prod_{i,j} P_{ij}$ where P_{ij} is an open subgroup of $E_{i,K_{ij}}$ for each i and j. Thus to complete the proof of Theorem 5.2.2, it will suffice to establish

Proposition 5.2.6 *Let K be a finite extension field of $\mathbb{Q}_p$ and E an absolutely simple algebraic group defined over K. Let w be a non-trivial word and P an open subgroup of E_K. Then $w(P)$ is open in P.*

Proof. This is an easy consequence of Theorem 5.1.4, which says that the morphism $f : E^{(k)} \to E$ given by evaluating w is *dominant* (i.e. its image is Zariski-dense). Now Corollary 1 of [PR], §3.1 says that in this case, if X is any non-empty open subset of $E_K^{(k)}$ then $f(X)$ contains a non-empty open subset of

E_K (being connected algebraic groups, E and $E^{(k)}$ are smooth and irreducible as algebraic varieties). In particular, it follows that $f(P^{(k)})$ contains a nonempty open subset U of E_K. Thus $w(P) \supseteq f(P^{(k)}) \supseteq U$ and it follows that $w(P)$ is open (see the next *exercise*).

A more elementary proof goes as follows. The p-adic analytic group P contains a torsion-free open pro-p subgroup H, say (see the proof of Theorem 5.3.3), and the same theorem shows that H is *just-infinite*, i.e. every non-identity closed normal subgroup of H is open. Now Corollary 5.3.2, proved below, says that in a non-soluble just-infinite pro-p group, *every* non-identity normal subgroup is open. So if we know (a) that H is non-soluble and (b) that $w(H) \neq 1$, we may conclude that $w(H)$ is open in H, and hence that $w(P)$ is open in P.

As E is a simple algebraic group it is not virtually soluble; it then follows by Platonov's theorem ([W1], **10.15**) that $w(E) \neq 1$. So both (a) and (b) follow from (c): H *is Zariski-dense in* E.

The proof of (c) is a nice interplay between the p-adic and Zariski topologies. It goes like this. Let M be the identity connected component of the Zariski-closure $\widetilde{H}$ of H in E. If $a \in E_K$ then H^a is commensurable with H, because $D := H \cap H^a$ is an open subgroup in each of the compact groups H and H^a. Therefore $\widetilde{D}$ is a Zariski-closed subgroup of finite index in both $\widetilde{H}$ and $\widetilde{H}^a$, and it follows that the identity component of $\widetilde{D}$ is equal to both M and M^a. Thus E_K normalizes M. But E_K is Zariski-dense in E, by [PR] Lemma 3.2, so E normalizes M. Therefore $M = E$ and so $\widetilde{H} = E$ as claimed. ∎

Exercise. Let H be a topological group, W a subgroup of H and U a non-empty open subset of H. Show that if $W \supseteq U$ then W is open. [*Hint*: observe that $W = \bigcup_{h \in W} Uv^{-1}h$ for some $v \in U$.]

From a group-theoretic point of view, Theorem 5.2.2 is interesting because it may be applied to the profinite completions of many S-*arithmetic groups*. An S-arithmetic subgroup of G is a subgroup of $G_\mathbb{Q}$ that is commensurable with $G_{\mathbb{Z}_S}$, where

$$\mathbb{Z}_S = \mathbb{Z}[1/p \mid p \in S] \subset \mathbb{Q}.$$

Now $G_\mathbb{Q}$ may be naturally identified with a subgroup of G_{A_S}, via the inclusions $\mathbb{Q} \subset \mathbb{Q}_p$, and then $G_{\mathbb{Z}_S} \le G_\infty(S)$. If G has the *strong approximation property* with respect to S then $G_{\mathbb{Z}_S}$ is dense in $G_\infty(S)$; if Γ is an S-arithmetic subgroup of G it follows that the closure $\overline{\Gamma}$ of Γ is commensurable with $G_\infty(S)$. On the other hand, if G has the S-*congruence subgroup property* then the adelic topology, restricted to Γ, coincides with the *profinite* topology on Γ (in simple terms, this means that every subgroup of finite index in Γ contains a principal congruence subgroup). In this case, $\overline{\Gamma}$ is isomorphic to the profinite completion $\widehat{\Gamma}$ of Γ. Thus $\widehat{\Gamma}$ is a profinite form of G provided G has both strong approximation and the congruence subgroup property.

For strong approximation, see Chapter 7 of [PR], and for some introductory discussion see Window 9 in [LS3]. If G is a connected $\mathbb{Q}$-simple group, then G

has strong approximation with respect to S if and only if G is simply connected and $G_{\mathbb{Z}_S}$ is infinite. The S-congruence subgroup property holds in many cases, for example whenever G is a classical group of split type and (Lie) rank at least $2 - |S|$; see [PR], §9.5.

Suppose for example that $G = \mathrm{SL}_n$, with $n \geq 2$ and $n + |S| \geq 3$. Then $\widehat{\Gamma}$ is a profinite form of G whenever Γ is an S-arithmetic subgroup of G (specific examples: $\widehat{\mathrm{SL}_3(\mathbb{Z})}$, $\widehat{\mathrm{SL}_2(\mathbb{Z}[\frac{1}{2}])}$, etc.).

I have said nothing about the verbal properties of arithmetic groups themselves. This is a largely unexplored topic. However, there are many interesting results about the special case of *commutator width* (i.e. the width of γ_2), and it would be interesting to explore to what extent they generalize to other words. For example, Dennis and Vaserstein [DV] show that the group $\mathrm{SL}_n(R)$ has commutator width ≤ 6 if n is large and R is a Dedekind ring of arithmetic type, while it has infinite commutator width if $n > 2$ and $R = F[X]$ with F a field of infinite transcendence degree. (A key observation is that every elementary matrix is a commutator if $n \geq 3$.)

5.3   On just-infinite profinite groups

Much of the previous chapter was concerned with finding conditions for verbal subgroups in profinite groups to be closed. In some profinite groups this happens automatically, for the silly reason that *all* normal subgroups are closed.

A profinite group G is *just-infinite* if G is infinite but every proper (continuous) quotient of G is finite; in other words, if every non-identity closed normal subgroup of G is open.

Theorem 5.3.1 *Let G be a finitely generated just-infinite prosoluble group. If G is not soluble then every non-identity normal subgroup of G is open.*

This was proved for pro-p groups by Andrei Jaikin. Note that if G is a just-infinite pro-p group then the Frattini subgroup $\Phi(G)$ is open, so G is finitely generated. Thus we have

Corollary 5.3.2 [J2] *In a non-soluble just-infinite pro-p group every non-identity normal subgroup of G is open.*

To prove the theorem, let $1 \neq N \lhd G$ and set $T = \overline{N}$ (the closure of N). Then T is open in G so T is again a finitely generated prosoluble group; it follows by Corollary 4.7.3 that T' is closed. If G is not soluble T can't be abelian, so T' is open in G. Therefore NT' is also open, so NT' contains T and therefore $NT' = T$. It follows by Exercise 4.7.6 that $N = T$. So N is open in G.

(This simple argument, taken from [J2], quotes some rather difficult results from §4.7; if G is pronilpotent all one needs is the elementary Corollary 4.3.2.)

Being just-infinite is of course a very special – one might say unnatural – condition. In fact it arises quite naturally as an analogue of simplicity. An infinite profinite group can't be simple, and just-infinite is the next best thing. Quite concretely, simple algebraic groups tend to have just-infinite subgroups; this holds for arithmetic groups such as $\mathrm{PSL}_n(\mathbb{Z})$ for $n > 2$ (with the obvious definition of 'just-infinite' for abstract groups), and for the profinite groups occurring as follows:

Theorem 5.3.3 *Let K be a finite extension field of $\mathbb{Q}_p$ and E an absolutely simple algebraic group defined over K. Then every torsion-free compact open subgroup of E_K is just-infinite.*

Proof. The group $G = E_K$ is a Zariski-closed subgroup of $\mathrm{SL}_n(K)$ for some n, and $\mathrm{SL}_n(K)$ may be identified with a Zariski-closed subgroup of $\mathrm{SL}_{nr}(\mathbb{Q}_p)$ where $(K : \mathbb{Q}_p) = r$. Therefore G is a p-adic analytic group. Associated to this group are two Lie algebras: the $\mathbb{Q}_p$-Lie algebra $\mathcal{L}(G)$ defined in [DDMS], Chapter 9, and the $\overline{K}$-Lie algebra L of the algebraic group E, which is defined over K (here $\overline{K}$ denotes the algebraic closure of K). According to [PR], Lemma 3.1, we have

$$\mathcal{L}(G) = L_K$$

where L_K denotes the K-points of L.

The correspondence between algebraic groups and Lie algebras shows that L is a simple Lie algebra. It follows that L_K is simple as a Lie algebra over $\mathbb{Q}_p$; for if $I \neq 0$ is an ideal of L_K considered as a $\mathbb{Q}_p$-algebra then $[I, L]$ is a non-zero ideal of the $\overline{K}$-Lie algebra L, whence $L = [I, L] \subseteq I$. Thus $\mathcal{L}(G)$ is a simple $\mathbb{Q}_p$-Lie algebra.

Now let H be a torsion-free compact open subgroup of G and $N \neq 1$ a closed normal subgroup of H. Then H has an open normal subgroup P that is a uniform pro-p group ([DDMS], Chapter 8); and $N \cap P \neq 1$, because $|N : N \cap P| \leq |H : P| < \infty$. Let $M/(N \cap P)$ be the torsion subgroup of $P/(N \cap P)$. Then M is a uniform normal subgroup of P and the quotient P/M is uniform.

Each uniform group U has an intrinsic $\mathbb{Z}_p$-Lie algebra structure, which we will denote U_L ([DDMS], Chapter 4). *Loc. cit.* Proposition 4.31 shows that M_L is an ideal of P_L. On the other hand, $\mathbb{Q}_p \otimes_{\mathbb{Z}_p} P_L$ is isomorphic to the simple Lie algebra $\mathcal{L}(G)$ (*loc. cit.*, Chapter 9), so $\mathbb{Q}_p \otimes_{\mathbb{Z}_p} M_L = \mathbb{Q}_p \otimes_{\mathbb{Z}_p} P_L$. As $P_L/M_L \cong (P/M)_L$ is free as a $\mathbb{Z}_p$-module it follows that $M_L = P_L$. Therefore $M = P$ and we conclude that

$$|H : N| \leq |H : P| \cdot |P : M| \cdot |M : N \cap P| < \infty.$$

The result follows. ■

The book [KLP] contains an exhaustive discussion of p-adic analytic just-infinite pro-p groups, all of which essentially arise in this way.

5.4 p-adic analytic groups

We have seen in §4.3 that some words (the $J(p)$-words) are elliptic in all finitely generated pro-p groups, and some are not. Here we consider a special class of pro-p groups, those of *finite rank*. Properties of these groups are described in the book [DDMS], where it is shown that the following are equivalent for a topological group G and a prime p:

1. G is profinite and virtually a pro-p group of finite rank;

2. G is profinite, finitely generated (topologically) and virtually a powerful pro-p group;

3. G is isomorphic to a closed subgroup of $\mathrm{GL}_n(\mathbb{Z}_p)$ for some n;

4. G is a compact p-adic analytic group.

Theorem 5.4.1 [J1] *Every compact p-adic analytic group is verbally elliptic.*

This generalizes a weaker form of Proposition 5.2.6: the group E_K in that proposition is p-adic analytic, so if P is a compact open subgroup of E_K the theorem shows that $w(P)$ is closed. This is not as strong as saying that $w(P)$ is open, which of course will not be true in general for a compact p-adic analytic group P.

Let w be a word and G a compact p-adic analytic group. Then G has an open normal subgroup H which is a powerful pro-p group, of finite rank r, say. Let $s = |G : H|$ and put $W = w(G)$. We have to show that W is closed in G (Proposition 4.1.2), and begin by making some reductions.

1. $W = (W \cap H) \cdot G_w^{*s}$ (Lemma 1.1.2).

2. Let $B = \overline{W \cap H}$. Then $B = (W \cap H)\Phi(B)$, so

$$B = \langle y_1, \ldots, y_r \rangle \, \Phi(B) = \overline{\langle y_1, \ldots, y_r \rangle}$$

for some $y_1, \ldots, y_r \in W$. It follows (Corollary 4.3.2) that

$$B' = \overline{B'} = \prod_{i=1}^{r} [B, y_i] \leq W.$$

Thus B' is a closed normal subgroup of G contained in W; so replacing G by G/B' we may assume henceforth that $W \cap H$ is *abelian*.

3. Let T be the torsion subgroup of H. According to [DDMS], Theorem 4.20, T is finite and H/T is a uniform pro-p group (in particular, H/T is torsion-free). Suppose we show that WT/T is closed in G/T. Then $W \cap H$ is a subgroup of finite index in $WT \cap H$, which is a pro-p group of finite rank and closed in G. It follows that $W \cap H$ is open in $WT \cap H$ (*loc. cit.*, Theorem 1.17), and hence closed in G. Then W, being the union of finitely many cosets of $W \cap H$, is also closed in G. So replacing G by G/T we may assume that H *is a uniform pro-p group*.

Write $\mathbf{p} = p$ if p is odd, $\mathbf{p} = 4$ if $p = 2$, and set $H_1 = H^{\mathbf{p}}$. Let $\{t_1, \ldots, t_a\}$ be a transversal to the cosets of H_1 in G, and define a generalized word function ξ on H by

$$\xi = w'_{\mathbf{v}_1} \divideontimes \cdots \divideontimes w'_{\mathbf{v}_l}$$

where $\{\mathbf{v}_1, \ldots, \mathbf{v}_l\} = \{t_1, \ldots, t_a\}^{(k)}$ (so $l = a^k$; here k is the number of variables in w). According to Lemma 1.1.1, we have

$$w'_G(H_1) = \langle w'_{\mathbf{v}_1}(H_1), \ldots, w'_{\mathbf{v}_l}(H_1) \rangle = \xi(H_1) \lhd G.$$

The key result is

Proposition 5.4.2 *Let H be a uniform pro-p group and ξ a generalized word function on H such that $\xi(H^{\mathbf{p}})$ is abelian. Then $\xi(H^{\mathbf{p}})$ is closed in H.*

Accepting this for now, we can conclude the proof of the theorem. The proposition shows that $w'_G(H_1) = \xi(H_1)$ is closed in H, and hence also in G. It is also normal in G and contained in W. So replacing G by $G/w'_G(H_1)$ we reduce to the case where $H_1 \leq w^*(G)$, in which case G_w is a finite set. As G is a linear group in characteristic zero, w is concise in G (Theorem 1.4.2). Therefore $w(G)$ is finite, and hence closed.

Proof of Proposition 5.4.2. In §4.3 of [DDMS] it is explained how the uniform pro-p group H naturally has the structure of a free $\mathbb{Z}_p$-module of finite rank. Choosing a basis $\{a_1, \ldots, a_d\}$ we have $H = a_1\mathbb{Z}_p \oplus \cdots \oplus a_d\mathbb{Z}_p$, and we associate to each element $h \in H$ its co-ordinates $h(1), \ldots, h(d) \in \mathbb{Z}_p$ where

$$h = \sum_{i=1}^{d} a_i h(i).$$

Every automorphism of H is $\mathbb{Z}_p$-linear with respect to this structure; so if $\alpha \in \mathrm{Aut}(H)$ we have

$$h^{\alpha}(i) = \sum_{j=1}^{d} c_{ij} h(j) \quad (\forall h \in H) \tag{5.3}$$

where $(c_{ij}) \in \mathrm{GL}_n(\mathbb{Z}_p)$. (For all this, see [DDMS], §4.3.)

Corollary 9.13 of [DDMS] says that the co-ordinate system specified above makes $H^{\mathbf{p}} = H_1$ into a 'standard group'; this means that the group operations are given in terms of the co-ordinates by power series over $\mathbb{Z}_p$ (all such series converge because $H_1 = \mathbf{p}H$ and so the co-ordinates are all divisible by p). More generally, the mapping obtained by evaluating any fixed word on $H_1^{(k)}$ is likewise given by such power series; this is the content of *loc.cit.*, Lemma 8.25. Using (5.3), it is easy to adapt the proof to show, even more generally, that the same holds for any generalized word function.

Now let ξ be as in the statement of the proposition, put $B = \overline{\xi(H_1)}$ and let A/B be the torsion submodule of H/B (considering H as a $\mathbb{Z}_p$-module as above). Then A is a direct summand of H, and we are free to choose our $\mathbb{Z}_p$-basis so that

$$\xi(H_1) \leq A = a_1 \mathbb{Z}_p \oplus \cdots \oplus a_r \mathbb{Z}_p$$

for some $r \leq d$. Then there exist power series $f_1, \ldots, f_r \in \mathbb{Z}_p[[X_{11}, \ldots, X_{1d}, \ldots, X_{k1}, \ldots, X_{kd}]]$ such that

$$\xi(h_1, \ldots, h_k) = \sum_{i=1}^{r} a_i f_i (h_1(1), \ldots, h_1(d), \ldots, h_k(1), \ldots, h_k(d))$$

for all $h_1, \ldots, h_k \in H_1$.

The r-tuple $(f_1, \ldots, f_r)$ defines a mapping $F : \mathbf{p}\mathbb{Z}_p^{(kd)} \to \mathbb{Q}_p^{(r)}$; thus F is the same as ξ if we identify H_1 with $\mathbf{p}\mathbb{Z}_p^{(d)}$ and $H_1 \cap A = \mathbf{p}A$ with $\mathbf{p}\mathbb{Z}_p^{(r)} \subset \mathbb{Q}_p^{(r)}$. Then the additive group generated by $\mathrm{Im}(F)$ corresponds to $\xi(H_1)$, which generates the $\mathbb{Z}_p$-module B, so $\mathrm{Im}(F)$ spans $\mathbb{Q}_p^{(r)}$ as a $\mathbb{Q}_p$-vector space. Now the following theorem will be proved in the next section:

Theorem 5.5.4 *Let $f_1, \ldots, f_r$ be power series in m variables over $\mathbb{Z}_p$, let $e \geq 1$ and define $F : p^e \mathbb{Z}_p^{(m)} \to \mathbb{Z}_p^{(r)}$ by $F(\mathbf{a}) = (f_1(\mathbf{a}), \ldots, f_r(\mathbf{a}))$. Suppose that $F(\mathbf{0}) = \mathbf{0}$ and that $\mathrm{Im}(F)$ spans $\mathbb{Q}_p^{(r)}$ as a $\mathbb{Q}_p$-vector space. Then there exist $\mathbf{b}_1, \ldots, \mathbf{b}_r \in p^e \mathbb{Z}_p^{(m)}$ such that the mapping*

$$\widetilde{F} : p^e \mathbb{Z}_p^{(rm)} \to \mathbb{Z}_p^{(r)}$$

$$(\mathbf{a}_1, \ldots, \mathbf{a}_r) \mapsto \sum_{i=1}^{r} (F(\mathbf{a}_i + \mathbf{b}_i) - F(\mathbf{b}_i))$$

is open at 0.

This means that the image under $\widetilde{F}$ of any neighbourhood of 0 contains a neighbourhood of 0 in $\mathbb{Q}_p^{(r)}$. Applying this to our mapping F defined above, with $p^e = \mathbf{p}$, we infer that the additive group generated by $F(\mathbf{p}\mathbb{Z}_p^{(kd)})$ contains a non-empty open subset of $\mathbf{p}\mathbb{Z}_p^{(r)}$, and hence that $\xi(H_1)$ contains a non-empty open subset of A.

Therefore $\xi(H_1)$ is open in A, so closed in H, and the proposition is proved.

This beautiful proof of Theorem 5.4.1, due to Andrei Jaikin, is essentially an analytic version of Theorem 5.1.1. There is a quite different way of viewing this theorem, namely as a pro-p analogue of Corollary 2.6.2. Here is a sketch of an alternative proof, using more group theory and less analysis; it is based on an unpublished Oxford thesis by Nick Simons.

As above, let G be a profinite group having an open normal subgroup H that is a pro-p group of finite rank, and let w be a non-trivial word. In view of reduction step **2**, above, we may assume that $w(G) \cap H$ is abelian. Then

$w(H)' = 1$. Now H is a linear group over $\mathbb{Z}_p$; it follows by Platonov's theorem ([W1], **10.15**) that H is virtually soluble.

From this we may deduce that G has a closed normal subgroup N which is a torsion-free nilpotent pro-p group of finite rank such that G/N is virtually abelian. Now we have the following analogue of Theorem 2.5.1:

Theorem 5.4.3 *Let G be a profinite group and N a closed normal subgroup such that G/N is virtually nilpotent. Suppose that N is a torsion-free nilpotent pro-p group of finite rank. Then there exist $n \in \mathbb{N}$ and a closed virtually-nilpotent subgroup C of $N^{1/n}G$ such that $N^{1/n}G = N^{1/n}C$.*

Various points here require elaboration:

1. The groups $N^{1/n}$ and $N^{1/n}G$ are defined as abstract groups in §2.4. In the present case, $\left|N^{1/n} : N\right|$ is finite by Exercise 2.4.2. So $\left|N^{1/n}G : G\right|$ is also finite. Then $N^{1/n}G$ becomes a profinite group if we take the open subgroups of G as a base for the neighbourhoods of 1, and $N^{1/n}$ is a closed normal subgroup.

2. The theorem is proved in the same way as 'Case 1' of Theorem 2.5.1. This depends in turn on Propositions 2.5.4 and 2.5.5, which concern group cohomology with coefficients in a module M which has finite total rank as a $\mathbb{Z}$-module; in fact the same results hold when M is a module of finite total rank over any principal ideal domain ([LR], §10.3), and we apply them now to a free $\mathbb{Z}_p$-module of finite rank. The propositions were used in Lemma 2.5.7 to produce a virtually-nilpotent supplement L for an abelian normal subgroup; in the profinite situation, we simply replace L by its closure to obtain a closed virtually-nilpotent supplement. The argument then proceeds as before.

(Thus we don't have to worry about continuous cohomology. It is likely that the same results do hold for this theory; indeed it is not hard to see that if G is a strongly complete profinite group – in particular if G is virtually pro-p of finite rank – and M is a profinite G-module then *every* 1-cocycle of G in M is continuous. Whether every 2-cocycle is equivalent to a continuous one when G is a pro-p group of finite rank seems to be an interesting open problem. See [S12], where this is discussed for the special case of trivial modules.)

The proof of Theorem 5.4.1 then follows that of Theorem 2.6.1; in place of Lemma 2.6.3 one uses the easy fact that if H is a pro-p group of rank r (topologically) generated by a set S then H is (topologically) generated by r elements of S. Details are left as an *Exercise*.

In earlier sections, we established ellipticity results for profinite groups G by proving uniformity results for *finite* groups: a word w has finite width in G if and only if it has bounded width in all the finite groups comprising $\mathcal{F}(G)$. What does Theorem 5.4.1 say about finite groups? If G is a pro-p group of finite rank r, say, the family $\mathcal{F}(G)$ consists of finite p-groups of rank at most r. But we can't infer from this that a word w is uniformly elliptic in *all* finite p-groups of rank at most r, because there isn't a single pro-p group of finite rank that has all of these as images (see the *Exercise* below).

However, the theorem does suggest that one should seek a result along these lines:

Problem 5.4.1 Is every word uniformly elliptic in the class of all finite p-groups of rank r? If not, what about the class of powerful finite p-groups?

(Note that $\mathrm{rk}(G) = \mathrm{d}(G)$ if G is a powerful finite p-group; see [DDMS], Chapter 2.) A positive solution to this problem (either part), proved by finite group theory, would provide a third proof for Theorem 5.4.1.

To state a precise 'finite' analogue of the theorem, we would need to solve the following general problem (which is interesting in its own right):

Problem 5.4.2 Find conditions on a family $\mathcal{X}$ of finite p-groups that are necessary and sufficient for $\mathcal{X}$ to be contained in $\mathcal{F}(G)$ for some pro-p group G of finite rank.

Exercise (i) Let $m, n \in \mathbb{N}$. Construct a powerful finite p-group P_{mn} of rank 2 such that $|\gamma_n(P_{mn}) : \gamma_{n+1}(P_{mn})| = p^m$. [*Hint*: let $V_m = 1 + p^m \mathbb{Z}_p$ act by multiplication on $\mathbb{Z}_p$, and consider a suitable quotient of $\mathbb{Z}_p \rtimes V_m$.]

(ii) Suppose that G is a finitely generated pro-p group that has every P_{mn} as a quotient. Prove that G has infinite rank. [*Hint*: show that $|\gamma_n(G) : \gamma_{n+1}(G)|$ is infinite for every n. Then show that this can't happen if G has finite rank (cf. [DDMS], Theorem 4.8).]

5.5   Analytic stuff

We consider formal power series of the form

$$f = f(\mathbf{x}) = \sum c(\mathbf{e}) x_1^{e_1} x_2^{e_2} \ldots x_m^{e_m}$$

where each $c(\mathbf{e}) \in \mathbb{Z}_p$ and $\mathbf{e} = (e_1, \ldots, e_m)$ ranges over m-tuples of non-negative integers. If $\mathbf{a} \in p^t \mathbb{Z}_p^{(m)}$ where $t \geq 1$ then the series obtained by substituting a_i for x_i for each i converges to a value $f(\mathbf{a}) \in p^t \mathbb{Z}_p$.

Lemma 5.5.1 ([DDMS], Corollary 6.50) *If there exists $t \geq 1$ such that $f(\mathbf{a}) = 0$ for all $\mathbf{a} \in p^t \mathbb{Z}_p^{(m)}$ then $f = 0$ (i.e. every coefficient $c(\mathbf{e})$ is zero).*

The partial derivatives $f_{(i)} = \frac{\partial f}{\partial x_i}$ are defined termwise by the usual formula. For $\mathbf{a} \in p \mathbb{Z}_p^{(m)}$ the *differential* $Df_{\mathbf{a}} : \mathbb{Q}_p^{(m)} \to \mathbb{Q}_p$ is the linear map

$$(f_{(1)}(\mathbf{a}), \ldots, f_{(m)}(\mathbf{a}))$$

$$(y_1, \ldots, y_m) \mapsto \sum_{i=1}^{m} f_{(i)}(\mathbf{a}) y_i.$$

The next lemma is a weak form of Taylor's Theorem:

Lemma 5.5.2 *If there exists $t \geq 1$ such that $Df_{\mathbf{a}} = 0$ for all $\mathbf{a} \in p^t \mathbb{Z}_p^{(m)}$ then f is a constant (i.e. every coefficient $c(\mathbf{e})$ with $\mathbf{e} \neq \mathbf{0}$ is zero).*

Proof. Assume without loss of generality that $c(\mathbf{0}) = 0$. Now we argue by induction on the number m of variables. Suppose first that $m = 1$, and that $f \neq 0$. Then

$$f(x) = c(s)x^s + c(s+1)x^{s+1} + \cdots$$

where $s \geq 1$ and $c(s) \neq 0$. Taking $a = p^n$ we have

$$Df_a \equiv sc(s)p^{(s-1)n} \pmod{p^{sn}}$$

so $Df_a \neq 0$ if n is so large that $p^n \nmid sc(s)$.

Now let $m > 1$ and suppose the result holds for fewer variables. We can write

$$f(x_1, \ldots, x_m) = g(x_1, \ldots, x_{m-1}) + \sum_{n=1}^{\infty} g_n(x_1, \ldots, x_{m-1})x_m^n.$$

Then $g_{(i)}(x_1, \ldots, x_{m-1}) = f_{(i)}(x_1, \ldots, x_{m-1}, 0)$ for $i \leq m-1$, so by the inductive hypothesis $g = 0$. Now fix $\mathbf{b} \in p^t \mathbb{Z}_p^{(m-1)}$. Then for $a \in p^t \mathbb{Z}_p$ we have

$$0 = f_{(m)}(\mathbf{b}, a) = \sum_{n=1}^{\infty} n g_n(\mathbf{b})a^{n-1}.$$

Lemma 5.5.1 now shows that $n g_n(\mathbf{b}) = 0$ for each n. As this holds for each $\mathbf{b}$, it follows by the same lemma that $g_n = 0$ for each n. Thus $f = 0$. ∎

Now let $f_1, \ldots, f_r$ be power series as above, and write $F = (f_1, \ldots, f_r)$. For $\mathbf{a} \in p\mathbb{Z}_p^{(m)}$ we denote by $DF_{\mathbf{a}}$ the linear map

$$(Df_{1,\mathbf{a}}, \ldots, Df_{r,\mathbf{a}})^T : \mathbb{Q}_p^{(m)} \to \mathbb{Q}_p^{(r)}$$

(T denotes transpose; the matrix of $DF_{\mathbf{a}}$ is the Jacobian $(\frac{\partial f_i}{\partial x_j})_{|\mathbf{a}}$). The following is a consequence of (a generalized form of) the Inverse Function Theorem; it can easily be proved by adapting the proof of [DDMS], Theorem 6.37; or see [S8], II, §3.9 'Submersions'.

Proposition 5.5.3 *Let F be as above and suppose that $F(\mathbf{0}) = \mathbf{0}$. If the rank of $DF_{\mathbf{0}}$ is r then the mapping $p\mathbb{Z}_p^{(m)} \to p\mathbb{Z}_p^{(r)}$ given by evaluating F is open at $\mathbf{0}$.*

We are now ready for

Theorem 5.5.4 [J1] *Let $f_1, \ldots, f_r$ be power series in m variables over $\mathbb{Z}_p$, let $e \geq 1$ and define $F : p^e \mathbb{Z}_p^{(m)} \to \mathbb{Z}_p^{(r)}$ by $F(\mathbf{a}) = (f_1(\mathbf{a}), \ldots, f_r(\mathbf{a}))$. Suppose that*

$F(\mathbf{0}) = \mathbf{0}$ *and that* $\text{Im}(F)$ *spans* $\mathbb{Q}_p^{(r)}$ *as a* $\mathbb{Q}_p$-*vector space. Then there exist* $\mathbf{b}_1, \dots, \mathbf{b}_r \in p^e \mathbb{Z}_p^{(m)}$ *such that the mapping*

$$\widetilde{F} : p^e \mathbb{Z}_p^{(rm)} \to \mathbb{Z}_p^{(r)}$$

$$(\mathbf{a}_1, \dots, \mathbf{a}_r) \mapsto \sum_{i=1}^{r} (F(\mathbf{a}_i + \mathbf{b}_i) - F(\mathbf{b}_i))$$

is open at 0.

Proof. Let

$$V = \sum_{\mathbf{a} \in p^e \mathbb{Z}_p^{(m)}} \text{Im}(DF_{\mathbf{a}}) \le \mathbb{Q}_p^{(r)}.$$

Supoose that V is a proper subspace of $\mathbb{Q}_p^{(r)}$. Then there exist $\lambda_1, \dots, \lambda_r \in \mathbb{Z}_p$, not all zero, such that $\langle \lambda, \mathbf{v} \rangle := \sum_{i=1}^{r} \lambda_i v_i = 0$ for all $(v_1, \dots, v_r) \in V$. Put $h = \sum_1^r \lambda_i f_i$; then

$$Dh_{\mathbf{a}} = \sum_{i=1}^{r} \lambda_i Df_{i,\mathbf{a}} = 0$$

for all $\mathbf{a} \in p^e \mathbb{Z}_p^{(m)}$, so in fact $h = 0$ by Lemma 5.5.2. But then

$$\langle \lambda, F(\mathbf{a}) \rangle = h(\mathbf{a}) = 0$$

for all $\mathbf{a} \in p^e \mathbb{Z}_p^{(m)}$, a contradiction since $\text{Im}(F)$ spans $\mathbb{Q}_p^{(r)}$.

It follows that $V = \mathbb{Q}_p^{(r)}$. Hence there exist $\mathbf{b}_1, \dots, \mathbf{b}_r \in p^e \mathbb{Z}_p^{(m)}$ such that

$$\text{Im}(DF_{\mathbf{b}_1}) + \cdots + \text{Im}(DF_{\mathbf{b}_r}) = \mathbb{Q}_p^{(r)}.$$

Now define power series g_i in rm variables by setting

$$g_i(\mathbf{x}_1, \dots, \mathbf{x}_r) = \sum_{j=1}^{r} (f_i(\mathbf{x}_j + \mathbf{b}_j) - f_i(\mathbf{b}_j))$$

$(i = 1, \dots, r)$. Then for $G = (g_1, \dots, g_r)$ the linear map $DG_{\mathbf{0}} : \mathbb{Q}_p^{(rm)} \to \mathbb{Q}_p^{(r)}$ is given by

$$(\mathbf{y}_1, \dots, \mathbf{y}_r) \mapsto \mathbf{y}_1 DF_{\mathbf{b}_1} + \cdots + \mathbf{y}_r DF_{\mathbf{b}_r},$$

and is therefore surjective. Thus $DG_{\mathbf{0}}$ has rank r, and so by Proposition 5.5.3 the mapping $\widetilde{F} : p^e \mathbb{Z}_p^{(rm)} \to p \mathbb{Z}_p^{(r)}$ defined by G is open at $\mathbf{0}$. ∎

Appendix: the problems

Stroud's theorem

Problem 1.2.1 Does γ_2 have finite width in every finitely generated soluble group?

This is Problem 1 in Stroud's thesis [S11], where he speculates that the answer is 'no'. His main result was that finitely generated abelian-by-nilpotent groups are verbally elliptic (Theorem 2.3.1). He (and independently Romankov) also showed that the 2-generator free group in the variety $\mathfrak{N}_2\mathfrak{A}$ (consisting of groups G with $\gamma_3(G') = 1$) is not verbally elliptic (see §3.2). The next two problems (which contradict each other) are aimed at locating the exact boundary between elliptic and non-elliptic varieties of soluble groups.

Problem 2.3.1 *Find a finitely generated centre-by-metabelian group that is not verbally elliptic.*

Problem 2.3.2 *Prove that every finitely generated group G such that*

$$[\gamma_n(G)',_m G] = 1$$

for some $n, m \in \mathbb{N}$ is verbally elliptic.

Finite groups

The 'Key Theorem' of Nikolov and Segal (Theorem 4.7.1) comes in three flavours, A, B and C. Each part has an undesirable feature in its hypothesis or its conclusion. A positive answer to the first problem would sweep these away:

Problem 4.7.1 Can $f_4(d, s)$ be made independent of s in Theorem 4.7.1(C)?

If the answer is yes, it implies a positive solution to the next problem, which is one of the major challenges in the subject:

Problem 4.7.2 Is G^q closed in G for every finitely generated profinite group G? Equivalently, is the word x^q uniformly elliptic in finite groups?

A word w is d-*restricted* if $|\mathbb{Z} : w(\mathbb{Z})|$ is finite and there exists a finite δ such that for every finite d-generator group G, the subgroup $w(G)$ is generated by δ w-values. Every d-locally finite word has this property; it seems natural to ask:

Problem 4.7.4 Is every d-restricted word d-locally finite?

More important is

Problem 4.7.3 For which natural numbers q and d is the word x^q d-restricted?

It follows from Corollary 4.7.11 that these are precisely the pairs (q, d) such that G^q is closed in G for every d-generator profinite group G, so the solution of this problem solves Problem 4.7.2.

Added in proof: As this book goes to press, we have just found a proof for the following

Theorem *The word* $\beta(q) = x^q$ *is* d-*restricted for every* q *and* d.

See [NS4]. This turns out to be a surprisingly simple consequence of Theorem 4.7.5, combined with Proposition 10.1 of [NS1]. It implies a positive answer to Problem 4.7.2 and a negative answer to Problem 4.7.4. It also implies that the d-restricted words are precisely those that are not commutator words (i.e. do not lie in the derived group of the free group on their variables); with Theorem 4.7.9 this yields the

Theorem *Every non-commutator word is uniformly elliptic in finite groups.*

Uniformly elliptic words

The first problem focuses on an obstacle to proving that all J-words are uniformly elliptic in finite soluble groups:

Problem 4.8.1 Does there exist a prosoluble group H having a normal subgroup K such that H/K is perfect?

Note that H can't be finitely generated, by Exercise 4.7.6, and that H/K has no proper subgroups of finite index, by Exercise 4.2.6.

Problem 4.8.2 Let $w = [x, y, \ldots, y]$ be an 'Engel word'. Is the relatively free group $F_\infty/w(F_\infty)$ residually finite?

If the answer is 'yes', then w is uniformly elliptic in all finite groups.

Adelic groups

The following has been proposed by László Pyber:

Problem 5.2.1 Is every finitely generated closed subgroup of $\prod_p \mathrm{SL}_n(\mathbb{Z}_p)$ verbally elliptic?

The answer is 'yes' for groups having no infinite virtually abelian quotients (the so-called *FAb groups*) [S6]; I believe it is 'no' in general.

Finite p-groups

Problem 5.4.1 Is every word uniformly elliptic in the class of all finite p-groups of rank r? If not, what about the class of powerful finite p-groups?

This is suggested by (and would imply) Jaikin's theorem that pro-p groups of finite rank are verbally elliptic (Theorem 5.4.1).

The last problem is of more general interest:

Problem 5.4.2 Find conditions on a family $\mathcal{X}$ of finite p-groups that are necessary and sufficient for $\mathcal{X}$ to be contained in $\mathcal{F}(G)$ for some pro-p group G of finite rank.

Bibliography

[A]   M. Abért, On the probability of satisfying a word in a group, *J. Group Theory* **9** (2006), 685–694.

[AG]   M. Aschbacher and R. Guralnick, Some applications of the first cohomology group, *J. Algebra* **90** (1984), 446–460.

[BNP]   L. Babai, N. Nikolov and L. Pyber, Expansion and product decompositions of finite groups: variations on a theme of Gowers, *to appear.*

[B1]   A. Borel, On free subgroups of semi-simple groups, *L'Enseignement Math.* **29** (1983), 151–164.

[B2]   A. Borel, *Linear algebraic groups, 2nd ed.,* Springer-Verlag, New York, 1991.

[BM]   R. G. Burns and Y. Medvedev, Analytic relatively free pro-*p* groups, *J. Group Theory* **7** (2004), 533–541.

[BMM]   R. G. Burns, O. Macedońska and Y. Medvedev, Groups satisfying semigroup laws, and nilpotent-by-Burnside varieties, *J. Algebra* **195** (1997), 510–525.

[C]   D. Calegari, *Stable commutator length* (in preparation); available at `www.its.caltech.edu/~dannyc/scl/`

[DV]   R. K. Dennis and L. N. Vaserstein, On a question of M. Newman on the number of commutators, *J. Algebra* **118** (1988), 150–161.

[DDMS] J. D. Dixon, M. P. F. du Sautoy, A. Mann and D. Segal, *Analytic pro-p groups, 2nd ed.*, Cambridge University Press, Cambridge, 1999.

[G1]   K. M. George, *Verbal properties of certain groups*, PhD thesis, University of Cambridge, 1976.

[G2]   D. Gorenstein, *Finite groups*, Harper and Row, New York, 1968.

[GLS]   D. Gorenstein, R. Lyons and R. Solomon, The classification of the finite simple groups, no. 1, 2nd ed., *Math. Surveys and Monographs* **40.1**, Amer. Math. Soc., Providence, Rhode Island, 2000.

[G3]   V. E. Govorov, Graded algebras (Russian), *Mat. Zametki* **12** (1972), 197–204; English translation *Math. Notes* **12** (1972), 552–556.

[H1]   B. Hartley, Subgroups of finite index in profinite groups, *Math. Zeit.* **168** (1979), 71–76.

[H2]   J. E. Humphreys, *Linear algebraic groups,* Springer-Verlag, New York, 1975.

[HW]   G. H. Hardy and E. M. Wright, *Introduction to the theory of numbers, 5th ed.,* Oxford University Press, Oxford, 1979.

[I]   S. V. Ivanov, P. Hall's conjecture on the finiteness of verbal subgroups, *Izv. Vyssh. Uchem. Zaved.* **325** (1989), 60–70.

[J1]   A. Jaikin-Zapirain, On the verbal width of finitely generated pro-*p* groups, *Revista Mat. Iberoamericana* **24** (2008), 617–630.

[J2]   A. Jaikin-Zapirain, On linear just infinite pro-*p* groups, *J. Algebra* **255** (2002), 392–404.

[JL]   M. Jarden and A. Lubotzky, Elementary equivalence of profinite groups, *Bull. London Math. Soc.* **40** (2008), 887–896.

[J3]   G. A. Jones, Varieties and simple groups, *J. Austral. Math. Soc.* **17** (1974), 163–173.

[KLP] G. Klaas, C. R. Leedham-Green and W. Plesken, *Linear pro-p groups of finite width,* Lect. Notes in Math. **1674**, Springer-Verlag, Berlin, 1997.

[LS1] M. Larsen and A. Shalev, Word maps and Waring type problems, *J. Amer. Math. Soc.,* to appear.

[LR]   J. C. Lennox and D. J. S. Robinson, *The theory of infinite soluble groups,* Clarendon Press, Oxford, 2004.

[LS2] M. W. Liebeck and A. Shalev, Diameter of simple groups: sharp bounds and applications, *Annals of Math.* **154** (2001), 383–406.

[LS3] A. Lubotzky and D. Segal, *Subgroup growth,* Birkhäuser, Basel, 2003.

[M1]   A. I. Mal'cev, Nilpotent semigroups, *Uchen. Zap. Ivanovsk. Ped. Inst.* **4** (1953), 107–111.

[MZ]   C. Martinez and E. Zelmanov, Products of powers in finite simple groups, *Israel J. Math.* **96** (1996), 469–479.

[M2]   Ju. I. Merzljakov, Algebraic linear groups as full groups of automorphisms and the closure of their verbal subgroups (Russian; English summary), *Algebra i Logika Sem.* **6** (1967) no. 1, 83–94.

[M3]   Ju. I. Merzljakov, Verbal and marginal subgroups of linear groups (Russian), *Dokl. Akad. Nauk SSSR* **177** (1967), 1008–1011.

[N]   H. Neumann, *Varieties of groups,* Ergebnisse der Math. **37**, Springer-Verlag, Berlin, 1967.

[NSS] M. F. Newman, C. Schneider and A. Shalev, The entropy of graded algebras, *J. Algebra* **223** (2000), 85–100.

[NP]   N. Nikolov and L. Pyber, Product decompositions of quasi-random groups and a Jordan-type theorem, *to appear.*

[NS1] N. Nikolov and D. Segal, On finitely generated profinite groups, I: strong completeness and uniform bounds, *Annals of Math.* **165** (2007), 171–238.

[NS2] N. Nikolov and D. Segal, On finitely generated profinite groups, II: products in quasisimple groups, *Annals of Math.* **165** (2007), 239–273.

[NS3] N. Nikolov and D. Segal, A characterization of finite soluble groups, *Bull. London Math. Soc.* **39** (2007), 209–213.

[NS4] N. Nikolov and D. Segal, Products of powers in finite groups, *in preparation.*

[P]   D. I. Piontkovskii, Hilbert series and relations in algebras (Russian), *Izv. Ross. Akad. Nauk Ser. Mat.* **64** (2000), 205–219; English translation *Izv. Math.* **64** (2000), 1297–1311.

[PR]   V. P. Platonov and A. Rapinchuk, *Algebraic groups and number theory,* Academic Press, New York, 1994.

[R1]   A. H. Rhemtulla, A problem of bounded expressibility in free products, *Proc. Cambridge Philos. Soc.* **64** (1968), 573–584.

[RZ]   L. Ribes and P. A. Zalesskii, *Profinite groups,* Ergebnisse der Math. **40**, Springer-Verlag, New York, 2000.

[R2]   V. A. Romankov, Width of verbal subgroups in solvable groups, *Algebra i Logika* **21** (1982), 60–72 (Russian); *Algebra and Logic* **21** (1982), 41–49 (English).

[SW]   J. Saxl and J. S. Wilson, A note on powers in simple groups, *Math. Proc. Cambridge Philos. Soc.* **122** (1997), 91–94.

[S1]   D. Segal, A residual property of finitely generated abelian-by-nilpotent groups, *J. Algebra* **32** (1974), 389–399.

[S2]   D. Segal, On abelian-by-polycyclic groups, *J. London Math. Soc.* **11** (1975), 445–452.

[S3]   D. Segal, *Polycyclic groups,* Cambridge University Press, Cambridge, 1983.

[S4]   D. Segal, Closed subgroups of profinite groups, *Proc. London Math. Soc.* **81** (2000), 29–54.

[S5]   D. Segal, Variations on a theme of Burns and Medvedev, *Groups, geometry and dynamics* **1** (2007), 661–668.

[S6]   D. Segal, On verbal subgroups of adelic groups, *to appear*.

[S7]   Z. Sela, Diophantine geometry over groups and the elementary theory of free and hyperbolic groups, *Bull. Symbolic Logic* **9** (2003), 51–70.

[S8]   J-P. Serre, *Lie algebras and Lie groups*, W. A. Benjamin, New York, 1965; Lect. Notes in Math. **1500**, Springer-Verlag, Berlin, 1992.

[S9]   J-P. Serre, *Galois cohomology*, Springer, Berlin, 1997.

[S10]   A. Shalev, Word maps, conjugacy classes and a non-commutative Waring-type theorem, *Annals of Math.*, to appear.

[S11]   P. W. Stroud, *Topics in the theory of verbal subgroups*, PhD thesis, University of Cambridge, 1966.

[S12]   B. Sury, Central extensions of p-adic groups; a theorem of Tate, *Communications in Algebra* **21** (1993), 1203–1213.

[T]   J. Tits, Classification of algebraic semisimple groups, in '*Algebraic groups and discontinuous subgroups*', *Proc. Symp. Pure Math.* **9**, Amer. Math. Soc., 1966, 32–62.

[TS]   R. F. Turner-Smith, Finiteness conditions for verbal subgroups, *J. London Math. Soc.* **41** (1966), 166–176.

[W1]   B. A. F. Wehrfritz, *Infinite linear groups*, Springer-Verlag, Berlin, 1973.

[W2]   J. C. R. Wilson, On outer-commutator words, *Canadian J. Math.* **26** (1974), 608–620.

[W3]   J. S. Wilson, On simple pseudofinite groups. *J. London Math. Soc.* **51** (1995), 471–490.

[W4]   J. S. Wilson, *Profinite groups*, Clarendon Press, Oxford, 1998.

[Z1]   E. I. Zelmanov, The solution of the restricted Burnside problem for groups of odd exponent, *Math. USSR Izv.* **36** (1991), 41–60.

[Z2]   E. I. Zelmanov, The solution of the restricted Burnside problem for 2-groups, *Mat. Sb.* **182** (1991), 568–592.

Index